Software Defined Radio for 3G

For a listing of recent titles in the *Artech House Mobile Communications Series*, turn to the back of this book.

Software Defined Radio for 3G

Paul Burns

AH

Artech House
Boston • London
www.artechhouse.com

Library of Congress Cataloging-in-Publication Data
Burns, Paul (Paul Gowans), 1965–
    Software defined radio for 3G / Paul Burns.
        p. cm.—(Artech House mobile communications series)
    Includes bibliographical references and index.
    ISBN 1-58053-347-7 (alk. paper)
        I. Software radio. 2. Global system for mobile communications. I. Title. II. Series.

TK5103 4875 .B87 2002   CIP 2002032655
621.384—dc21

British Library Cataloguing in Publication Data
Burns, Paul
    Software defined radio for 3G.—(Artech House mobile communications series)
    I. Software radio. 2. Cellular telephone systems
    I. Title
    621.3'8456

    ISBN 1-58053-347-7

Cover design by Igor Valdman

MATLAB® and Simulink® are registered trademarks of The MathWorks, Inc.

© 2003 ARTECH HOUSE, INC.
685 Canton Street
Norwood, MA 02062

International Standard Book Number: 1-58053-347-7
Library of Congress Catalog Card Number: 2002032655

10 9 8 7 6 5 4 3 2 1

To my wife, Josephine, for her love, encouragement, enthusiasm, and sacrifices during the day-to-day development of this book.

To my father, Paul, (senior) for passing on a passion for learning and intellectual stimulation.

To my mother, Jean, for nuturing me through my youth and supporting my every venture.

To my brother, Kenneth, for our life-long friendship and common enjoyment of all things radio.

And to my brothers and sisters-in-law and the rest of my family for their moral support and interest in this project.

Contents

Preface

Software defined radio (SDR) is an exciting new field for the wireless industry; it is gaining momentum and beginning to be included in commercial and defense products. The technology offers the potential to revolutionize the way radios are designed, manufactured, deployed, and used. SDR promises to increase flexibility, extend hardware lifetime, lower costs, and reduce time to market.

In 1999, when I started to design a 3G software defined mobile cellular base station, there were no books on the subject; the only sources of information were dispersed across the Internet and journal papers. This book was written to help solve that problem and provide the reader with a complete text covering design issues, applications, and reference examples that can be applied to 3G and other applications.

This book is designed for engineering professionals who are considering software radio as a solution to a new product development, particularly if it is related to 3G cellular mobile and the next wave of standards. University course leaders and students can also use this book as the basis for the study of software radio, especially if they have a practical bias.

We look at systems design and partitioning all the way from the antenna to the management and control software. Various options for hardware are provided, including a look at current and state-of-the-art silicon technologies such as ADCs, DACs, DSPs, FPGAs, RCPs, ACMs, and digital frequency up and downconverters. Software radio is becoming popular in wireless communications, and this book covers both TDMA- and CDMA-based mobile cellular radio systems. The technology is an enabler for many new capacity improving applications including smart antennas and multiuser detection. A chapter detailing the actual implementation of a low-cost software radio using off-the-shelf components gives readers a great headstart to the world of software radio.

Chapter Summary

Chapter 1: What Software Defined Radio Is and Why We Should Use It

This chapter is an introduction to the concepts of software defined radio, signal processing, and 3G. The differences between a hardware and a software radio are explained and some history about digital signal processing and the SDRF is provided.

Chapter 2: A Basic Software Defined Radio Architecture

Chapter 2 is an introduction to software defined radio architectures and the presentation of a reference SDR block diagram. This chapter introduces the issue of system-level partitioning and estimates the signal processing load required for the digital frequency conversion function. Baseband signal processing is introduced and the chapter concludes with an example COTS hardware architecture.

Chapter 3: RF System Design

Cellular mobile performance standards place rigorous requirements on radio receivers and transmitters. Multiple air interface systems face conflicting requirements when attempting to cover both TDMA and CDMA air interfaces. This chapter explores the basics of communications theory and provides the equations for the IMT 2000 path loss models. The reader is provided with a complete link budget, which can be used when designing the analog RF components of a 3G software radio—for example, power amplifiers, mixers, low noise amplifiers, and so on. A major section is devoted to detailing 3G performance requirements for such parameters as dynamic range, blocking, and intermodulation. Important software radio RF components such as multicarrier power amplifiers are introduced and techniques for linearization are detailed. The chapter concludes with a proposed software radio design flow with special emphasis on the RF stages.

Chapter 4: Analog-to-Digital and Digital-to-Analog Conversion

Arguably the most critical function in a wideband software radio, we investigate the analog-to-digital conversion (ADC) and digital-to-analog conversion (DAC) performance requirements needed for 3G cellular systems. The chapter discusses the fundamentals of transforming between the analog and digital domains, including Nyquist's theorem, bandpass sampling, aliasing, and quantization. Performance parameters such as SFDR, dynamic range, and jitter are introduced. The architecture of a state-of-the-art A/D converter is provided, and techniques for achieving wideband SFDRs approaching 100 dB are detailed. A figure of merit (FOM) for software radio applications is provided as an initial guide when selecting a converter from the large range of commercially available options. The chapter concludes with an analysis of the allowable noise contributions from ADCs and DACs and discusses the tough GSM blocking specification.

Chapter 5: Digital Frequency Up- and Downconverters

For wireless systems operating above the 20–100-MHz range, digital frequency up and downconverters are currently the most efficient bridge between digital intermediate frequencies and baseband processors. This chapter introduces the concepts of frequency conversion, decimation, interpolation, multirate processing, NCOs, halfband filters, and CIC filters and provides the equations for IIR and FIR filters. A detailed overview of several commercially available digital frequency converters is provided along with a description of the components' functionality. Figures illustrating the 3G performance for these converters are also included.

Chapter 6: Signal Processing Hardware Components

An important goal for 3G software radio is to perform all symbol and chip-rate processing in software. This chapter commences with an analysis of the required signal processing power for UMTS and introduces the major CDMA layer one processing functions—for example, path searching and rake receivers. Architecture details of candidate hardware devices such as DSPs, FPGAs, RCPs, and ACMs are provided, together with an estimate for the number of users that devices can support. The chapter concludes with symbol and chip-rate partitioning.

Chapter 7: Software Architecture and Components

Reconfiguring the radio with software is the most important requirement, and this chapter starts by discussing the ideas of hardware abstraction and then proceeds with details of the software architectural proposals produced by the JTRS JPO and the SDRF. These specifications use object-oriented design principles; the chapter includes many UML figures to explain the designs.

Software communications protocols and operating systems such as CORBA and RT LINUX are discussed, and the chapter concludes with an overview of the languages used to program the hardware: C, C++, VHDL, and Verilog.

Chapter 8: Applications for Wireless Systems

To design an efficient and cost-effective 3G software radio it is necessary to have a depth of understanding that extends from the wireless network components (e.g., MSC, BSC, BTS, and terminal) down to the radio's air interface. This chapter commences by explaining the fundamental concepts behind CDMA transmission and reception and covers the rake receiver in more detail, as well as introducing the concepts of handover and power control. Special attention is given to the WCDMA/UMTS, CDMA2000, and GSM air interfaces, with explanations of techniques used in each for multiple access, modulation, and spreading. A major part of the chapter is devoted to several example software radio implementations, with details of the software and hardware architectures. The chapter concludes with details of an example 3G network in Korea, including statistics about the number of base stations, network coverage, and so on.

Chapter 9: Smart Antennas Using Software Radio

Software defined radio is an enabling technology, and smart antennas are one of the ideal applications. Architectures for implementing smart antennas on a software radio platform are introduced. There are many classes of candidate algorithms (e.g., statistically optimum, blind adaptive, and so on) available to perform the necessary processing, and the chapter summarizes these. Smart antenna processing is a viable approach for increasing user capacity but must be traded off for increased signal processing power; the chapter analyzes the WCDMA/UMTS case and estimates the processing load. The chapter concludes with a block diagram of a flexible hardware architecture suitable for implementing smart antenna software.

Chapter 10: Low-Cost Experimental Software Radio Platform

This chapter puts the theory of previous chapters into practice and provides the outline of a design for an experimental, low-cost software radio platform that can be purchased off-the-shelf. The platform has the capability to teach the fundamentals of software radio and is ideal for university labs and small-scale development. Software can be developed in C and makes use of a DSP operating system if required. The importance of memory management and FIFO buffers is stressed; sampling rate issues are covered, as well as many of the implementation-level details, including register settings.

Chapter 11: Engineering Design Assistance Tools

The complexity of current digital radio air interfaces almost mandates the use of engineering design assistance (EDA) tools during the development of new products; this applies equally well to software and hardware implementations. This chapter explores the benefits of EDA tools and reviews several popular tools well suited to software radio and 3G cellular systems development.

I hope you find this book helpful and that after reading the work you will be enthused enough to contribute to the advancement of this fantastic technology. Please feel free to contact me at http://www.simplexity.com.au.

1

What Software Defined Radio Is and Why We Should Use It

This book concentrates on cellular mobile radio and 3G in particular; however, the content can be applied to other radio applications. Software defined radio (SDR) is growing in popularity and the expression is becoming recognized in the wider technical community.

1.1 Introduction to Software Defined Radio

The twentieth century saw the explosion of hardware defined radio (HDR) as a means of communicating all forms of audible; visual, and machine-generated information over vast distances. Most radios are hardware defined with little or no software control; they are fixed in function for mostly consumer items for broadcast reception. They have a short life and are designed to be discarded and replaced.

Software defined radio uses programmable digital devices to perform the signal processing necessary to transmit and receive baseband information at radio frequency. Devices such as digital signal processors (DSPs) and field programmable gate arrays (FPGAs) use software to provide them with the required signal processing functionality. This technology offers greater flexibility and potentially longer product life, since the radio can be upgraded very cost effectively with software.

A major challenge for software defined radio is to equal the efficiencies of purely hardware solutions while providing the flexibility and intelligence that software can offer. Efficiencies can be measured by the cost per informa-

1

tion bit; the power consumed per information bit, and the physical volume consumed per information bit. Also, the user will not need or want to know the underlying technology of the radio but will still demand higher efficiency, more flexibility, and greater intelligence. At the same time the software radio applications developer will want to be shielded from the details of the computing and signal processing hardware and complete all development in a unified environment using a single high-level language.

The latest explosion in radio communications caused by the cellular mobile phone is a prime contributor to the effort being invested in SDR. Combined with a seemingly endless exponential growth in silicon chip computing power, the twenty-first century is sure to see radio communications expand and software radio play an increasingly significant role.

1.2 3G Software Radio Applications

The first generation of mobile cellular communications commenced in the 1980s and used analog modulation techniques to transmit and receive analog voice only information between mobiles and base stations. Second-generation (2G) systems of the early 1990s were known as "digital" because they encoded voice into digital streams and used digital modulation techniques for transmission. The digital nature of 2G allows limited fixed rate data services.

The International Telecommunication Union (ITU) developed the IMT 2000 standard to define the requirements for 3G-compatible systems. This standard includes such provisions as the ability to support up to 2 Mbps data connections. Some see 3G as a means to provide new services to customers, while others see the next generation as purely a means of providing much needed capacity via better spectrum utilization.

An important problem with 3G is that it has added to the number of air interface standards that needs to be supported by the infrastructure and terminal industry. Of the 3G standards, the 3GPP [1] Universal Mobile Telecommunications System (UMTS) is unlikely to be universal and will be strongest in Europe. The 3GPP2 [2] CDMA2000 standard and the TDMA-based GSM-EDGE systems will be successful in North and South America, while Japan has its own WCDMA system similar to UMTS.

All of the 3G systems are potential SDR applications. Software radio offers the potential to solve many of the problems caused by the proliferation of new air interfaces. Base stations and terminals using SDR architectures can support multiple air interfaces during periods of transition and be easily software upgraded. Intelligent SDRs can detect the local air interface and adapt to suit the need; this capability will be valuable for frequent intercountry travelers.

1.3   A Traditional Hardware Radio Architecture

To appreciate where software radio is heading it is first useful to review a traditional hardware radio architecture. Figure 1.1 illustrates a dual conversion superheterodyne transceiver. This design has been around since the 1930s and it is almost certain that a majority of homes would possess a superheterodyne receiver of some sort (broadcast radio, television, and so on).

From the receiver point of view the RF from the antenna is converted down to an intermediate frequency by mixing or multiplying the incoming signal with the first local oscillator, LO1. The IF is filtered and then mixed down to baseband by the second oscillator, LO2, and mixer. The baseband modulated signal is demodulated to produce the analog receive information, and the reciprocal functions are performed for the transmitter. The number of conversion stages is dependent upon the RF operating frequency, and theoretically it is possible to add stages and push the operating frequency higher. The analog superheterodyne radio has experienced a marvelously successful history; it was used in 1G mobile phone terminals and is sure to endure in low-cost broadcast radio receivers for many years to come. This architecture was suited to 1G mobile phone systems, such as advanced mobile phone system (AMPS), which used frequency modulation (FM) and frequency division multiplexing (FD) to allow multiple users to access a fixed piece of spectrum.

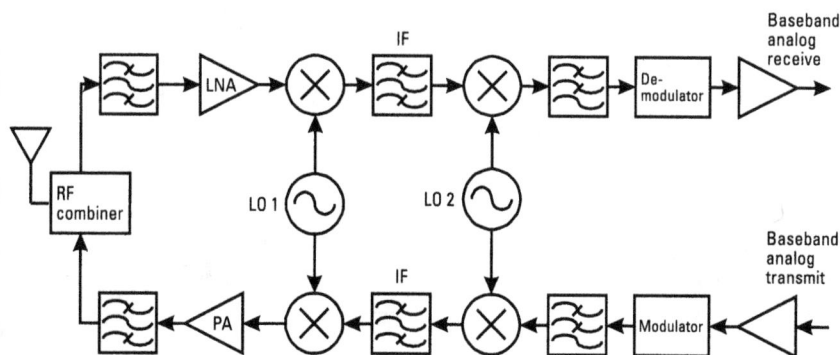

Figure 1.1   Traditional hardware radio architecture.

The AMPS system allocates a dedicated 30-kHz spectrum slice to each user irrespective of the amount of information required to be exchanged.

1.4 An Ideal Software Defined Radio Architecture

The ideal software radio architecture shown in Figure 1.2 consists of a digital subsystem and a simple analog subsystem. The analog functions are restricted to those that cannot be performed digitally—that is, antenna, RF filtering, RF combination, receive preamplification, transmit power amplification and reference frequency generation.

The architecture pushes the analog conversion stage right up as close as possible to the antenna, in this case prior to the power amplifier (PA) in the transmitter and after the low noise amplifier (LNA) in the receiver. The separation of carriers and up/down frequency conversion to baseband is performed by the digital processing resources. Similarly, the channel coding and modulation functions are performed digitally at baseband by the same processing resources.

Software for the ideal architecture is layered so that the hardware is completely abstracted away from the application software. A middleware layer achieves this functionality by wrapping up the hardware elements into objects and providing services that allow the objects to communicate with

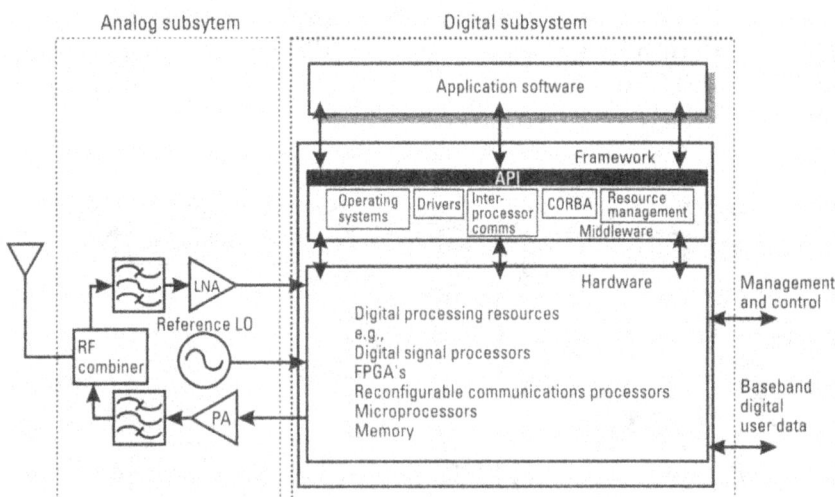

Figure 1.2 Ideal software defined radio with layered hardware and software.

each other via a standard interface—for example, Common Object Request Broker Architecture (CORBA). Middleware includes the operating system, hardware drivers, resource management, and other non-application-specific software. The combination of hardware and middleware is often termed a framework.

Future SDR designs and frameworks that use an open API into the middleware will make applications development more portable,quicker,and cheaper. Applications developers will be freed from designing ways to program the low-level hardware and allowed to concentrate on building more complicated and powerful applications.

The ideal architecture is commercially feasible for limited low data rate HF and VHF radios but is not yet practical for any generation of cellular mobile phone technology. The ideal architecture is useful as a point of comparison and acts as a guide for the development of hardware and middleware in the future. Practical architectures for cellular radio are covered in Chapters 2 and 8.

1.5 Signal Processing Hardware History

The birth of high-speed, cost-effective digital signal processing occurred in 1983 when Texas Instruments (TI) released the TMS320 chip, a 5 million

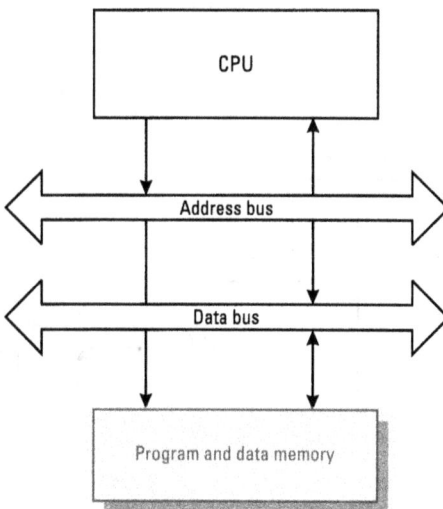

Figure 1.3   Von Neumann memory architecture.

instructions per second (MIPS) device that was faster than any comparable digital signal processor (DSP) of the time. The TMS320 has progressed through several generations of design, with the current TMS320C64X [3] claiming an impressive 1,760 times increase to 8,800 MIPS. TI's early success with the DSP was followed by several other companies, including Analog Devices, AT&T, and Motorola.

DSP chip designers in the 1980s realized that general-purpose computing architectures (e.g., Intel 386) could be improved upon to suit high-speed signal processing by providing the ability to load operands at the same time as instructions are fetched. Microcontrollers extensively use the Von Neumann memory architecture, as shown in Figure 1.3. The single data bus causes a bottleneck in the system by only allowing either new instructions or data to be fetched from external memory and loaded into the CPU.

Many DSP chips avoid the instruction and data contention by employing the Harvard architecture, as shown in Figure 1.4. By using two address buses and two data buses, each connected to its own piece of external memory, it is possible for new instructions to be fetched at the same time as new data. This allows for effective pipelining, where instructions for the next series of data can be loaded at the same time as operations are performed on the current set of data.

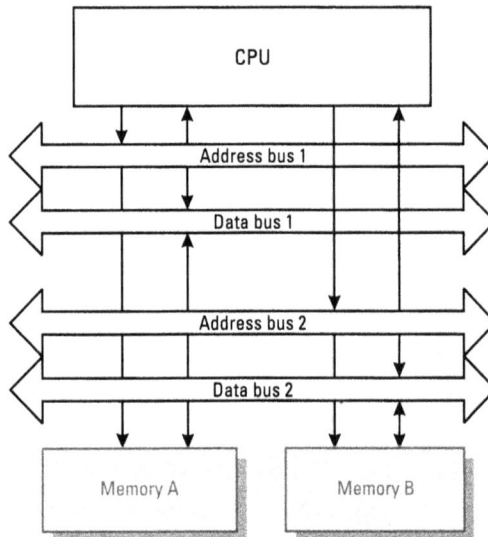

Figure 1.4   Harvard memory architecture.

The other big change with the development of a DSP-centric architecture is the ability to perform the multiply and accumulate functions in a single clock cycle. General-purpose processors without a dedicated multiplier require many shift and add operations to achieve the same result, consuming precious clock cycles. Many communications-related signal processing algorithms are both multiply and accumulate (MAC) intensive and repetitive, where a relatively small set of instructions is performed over and over in tight loops.

The DSP architecture based on Harvard memory design and single cycle MAC continues to evolve. The latest DSP hardware and programming techniques are specifically tailored for a range of new technologies, including 3G communications.

Signal processing hardware is covered in more detail in Chapter 6.

1.6 Software Defined Radio Project Complexity

Improving technology and reduced time to market pressures are increasing the complexity of hardware. These same demands and the specification of more bandwidth efficient air interfaces are stressing the ability of engineers to design and implement software within budgeted cost and schedules.

Software engineering practices have progressed from designing structured systems with pen and paper to object-oriented designs assisted by computer-aided tools. Some of the changes have been cosmetic, while most are the result of experiences learned from software project failures.

Almost any software radio project and certainly any 3G radio project will qualify as a large, complex, Software-intensive system. Large software system developments are very often underestimated, poorly planned, and inconsistently implemented. Special attention to planning and the careful selection of the correct engineering design assistance (EDA) tools are absolutely essential for a successful project.

1.7 The Software Defined Radio Forum

The Software Defined Radio Forum (formerly the Modular Multifunction Information Transfer Systems forum) was formed in 1996 as the result of U.S. government action to promote open standards architecture for SPEAKeasy, a military software radio project. The Forum describes itself as follows: The SDR Forum is an open, non-profit corporation dedicated to supporting the development, deployment, and use of open architectures for advanced wireless systems. The Forum membership is international, and growing [4].

Membership is represented by software radio manufacturers, telecommunications infrastructure and terminal manufacturers, silicon chip vendors, test equipment makers, telecommunications companies, scientific and research organizations, and others from the commercial and defense sector.

There are three parts to the Forum's technical committee: the download/handheld, base station/smart antennas, and mobile working groups. Each committee has its own aim with the handheld group aiming to promote the use of software defined radio technology in handheld terminals providing dynamic reconfiguration under severe constraints on size, weight, and power [4]. The base station committee is aiming to promote the use of software defined radio and reconfigurable adaptive processing technology in wireless base stations worldwide for terrestrial, satellite, mobile, and fixed services [4], while the mobile group seeks to promote the use of software defined radio technology in commercial and military applications under adverse terminal conditions where station mobility, dynamic networking, and operational flexibility are required using a variety of wireless and network interfaces [4].

The SDR Forum is not a standards body such as the TIA or ETSI; however, it does develop recommendations that may, in the future, turn into standards if enough commercial cooperation is developed. A range of documentation is made available on the Forum's Web site [4]; one of the most relevant is the Distributed Object Computing Software Radio Architecture recommendation. This work has come from the mobile working group and defines the software framework for an open, distributed, object-oriented, software programmable radio architecture. The proposal makes extensive use of CORBA to provide middleware services for the applications and hardware. Object concepts and CORBA are discussed in more detail in Chapter 7.

1.8 Conclusion

When an engineer or scientist needs to design a new radio should he or she be thinking SDR or HDR? The answer to this question, of course, depends upon the designer's radio application, flexibility needs, power consumption requirements, size constraints, organizational expertise, time scale, funding, and not least the target build cost for the new product. The following chapters are designed to help answer these questions.

SDR 3G cellular base stations are now possible and can match the efficiency performance of their HDR predecessors. Some SDR products are taking the open architecture road, while others remain closed and proprietary. Extending the 3G example, it is possible that any large and scalable radio application under 3 GHz will be a suitable candidate for a software implementation. In this context base stations are "large" in processing capacity and "scalable," while terminals are relatively "small" capacity items with little need to scale up and support more users. The dramatic telecommunications/technology stock market crash during 2000 and the huge license fees paid by European telecommunications companies for 3G spectrum are likely to place greater pressure on the cost of building new networks and develop a need to gain longer life from existing networks. If the SDR can outperform the HDR (particularly from a cost view) and simultaneously provide flexibility, future enhancement, and intelligence, the technology should be well placed to play a significant role.

Market forces, increasing numbers of object-oriented software engineers, and the impact of Moore's Law driving up digital processing performance are all contributing to a gathering momentum for SDR as a design choice. This momentum will spur innovation and drive cost lower so that over coming years SDR may become an imperative rather than a choice.

The largest potential driver for growth will be the adoption of an open architecture. Just as the open architecture IBM-based personal computer (PC) and the Windows operating system made the WinPC (PC and Windows) a ubiquitous environment for application developers, it is possible that SDR could follow the same path and enjoy exponential growth leveraged from common hardware and software platforms. 3G has the potential to benefit from a successful development of this approach and avoid the costs of reinvention that occurred during 2G, where every wireless vendor learned and paid for the same lessons.

In concluding this chapter it is hoped that SDR develops along a course that allows its maximum potential to be realized.

References

[1] http://www.3gpp2.org.

[2] http://www.3gpp.org.

[3] Texas Instruments, "TMS320C64X Technical Overview," SPRU395B (January 2001).

[4] http://www.sdrforum.org.

2

A Basic Software Defined Radio Architecture

In Chapter 1 we introduced the concept of software defined radio and explored some of the elements of technology that can be used during the design and development process. We discussed the hardware defined radio and then progressed forward to the ideal SDR, avoiding any coverage of the ground in-between. The ideal SDR will be used as a reference point throughout this book.

2.1 Introduction

For 3G mobile and many other multiuser radio technologies, the ideal SDR is not yet a practical or cost-effective reality. Direct sampling of wideband (100s of kHz to 10s of MHz) RF frequencies (100s of MHz to 10s of GHz) at high signal to noise ratio (>90 dB), as required by the ideal multicarrier SDR, is not yet technically possible. Direct RF sampling for single carrier terminal/mobile devices is becoming feasible; this area is explored in an article by Tsurumi and Suzuki [1].

So if we cannot use an ideal SDR, we must decide where the radio stops being hardware defined and where it starts being software defined. Design decisions, including the choice of sampling frequency, sampling bandwidth, and over or under/passband sampling, are key factors. The concept of where and how a design is started is an important one. Complex systems designs should start from the top down and the bottom up simultaneously, making their way to a conclusion via iteration. More detailed software design and process information is provided in Chapter 7.

This chapter focuses on bridging the gap between the HDR and the ideal SDR and exploring ways that the functional architecture can be successfully mapped onto practically available hardware. Since not every organization has the capability or desire to design and build high-speed digital hardware, an example implementation using commercial off-the-shelf (COTS) hardware is detailed. Considering normal commercial requirements (principally cost effectiveness), it is apparent that SDR implementations of 3G wireless need purpose-built hardware to be successful; examples of suitable hardware are provided in Chapter 8.

2.2 2G Radio Architectures

When compared with current generations, 1G and other equivalent analog radio systems trade off complexity for bandwidth utilization—that is, they are less complex and consume more bandwidth. 1G systems such as AMPS used a whole RF carrier per user and reduced spectrum use solely by applying the basic cellular concept and frequency reuse. Prior to 1G and the cellular concept, a user on a private mobile radio or early mobile phone system would occupy a whole RF carrier over an area the size of a complete city.

A major requirement during the development of the 2G standards was to increase bandwidth efficiency over that obtained with 1G; this resulted in a commensurate increase in system complexity. The 2G Groupe Speciale Mobile (GSM) standard achieved this by implementing a digital standard that allowed for time division multiplexing and other relatively sophisticated techniques, such as frequency hopping, discontinuous transmission, and power control. GSM multiplexes eight user channels onto a single carrier; the IS136 TDMA standard multiplexes three users. AMPS consumes 30 kHz for a voice user, while GSM occupies 200 kHz for its eight voice users. At first glance GSM only saves 40 kHz of bandwidth; however, it is the added features such as frequency hopping that provide for more efficient cell reuse patterns, producing an approximately three to four times capacity improvement.

The switch to a digital communications system for 2G provided other benefits and avenues for improvements in design. The multiple access component of the 2G standards reduced the number of carrier chains in base transceiver stations (BTSs). Both TDMA and CDMA multiplex a number of users in the digital part of the system at baseband prior to the modulator. For

GSM eight users are simultaneously frequency translated to the carrier by a single conversion chain, thereby saving the use of seven chains as would be required in a 1G system.

One downside for TDMA is that the high rate switching/multiplexing of the time domain transmission signals places stringent requirements on the linearity of the analog components following the modulator. Increased system linearity is necessary to ensure that spurious out of band emissions are kept to a level that allows excellent system capacity to be maintained. The linearity needed for 2G was achievable at the beginning of the 1990s for narrow band (single carrier) up and downconversion chains. There are indications that the specification limits for linearity set during the GSM standardization process were determined by what was achievable by the technology at the time, rather than considering actual need. It is possible that for many instances these linearity limits are too severe and could be relaxed to better suit SDR without any significant system capacity or performance issues.

Software radio architectures are characterized by large bandwidth generic frequency translation functions. High linearity and large bandwidth are competing parameters, and the performance requirements of TDMA systems stretch the software radio's ability to simultaneously achieve type approval and cost efficiency. This is not to say that the goals for software-based TDMA radio are now impossible, just that they were extremely difficult in the past and may have reduced the success for some early commercial software radio designs.

2.2.1 Hybrid Radio Architecture

We will now bridge the HDR to ideal SDR gap and explain the architectural progressions made by radio between HDR and the digital radios (2G) of the 1990s.

The analog fixed function HDR survived right through to the 1960s and 1970s, making its way into color television transmission, private mobile radio, and even parts of 1G cellular mobile radio. The complexity of a color television receiver and a 1G mobile terminal stretched this analog technology to the absolute limit. The very first color televisions and 1G cellular mobiles (that were more like portables) were very expensive and sometimes not that reliable. Analog electronic circuits consume more space and power and are more subject to performance variations as a result of environmental factors such as temperature changes.

The emergence of low-cost ADCs, DACs, and DSPs in the 1980s and the need for more efficient RF bandwidth utilization shifted radio architecture development away from purely analog to hybrid analog and digital systems. An example of a single RF carrier hybrid architecture is shown in Figure 2.1.

This architecture was popular with early 1990s digital radios, including those used in 2G BTSs. For a base station the receiver and transmitter typically occupy a different 20-MHz frequency segment. In GSM 900 the receiver can occupy a segment in the range 880–915 MHz while the transmitter uses a segment in the range 925–960 MHz. Because of the high operating frequency the system may require two or more intermediate frequencies (e.g., IF_{A1} and IF_{A2}) to achieve the required performance. The function of the analog section of the receiver is to select a single RF carrier from within the receive band and translate it to baseband. This is achieved by a series of conversion stages, each consisting of a multiplier and a filter. For the receiver the first conversion stage mixes the incoming band limited signal, F_{Rx}, with a tunable local oscillator, $LO1_{Rx}$, to select the required carrier and shift it to a common (to the system) analog intermediate frequency, IF_{A1}, typically 140 or 70 MHz. The filter in each conversion stage ensures that acceptable selectivity and image rejection are achieved. RF system considerations are covered in more detail in Chapter 3. The process is repeated by another downconversion stage to a second common intermediate frequency, IF_{A2}, typically 10.7 MHz. Finally, the last conversion stage mixes the required carrier to baseband or zero IF with a lowpass filter (LP) prior to the analog to digital converter.

The digitized baseband signal is then demodulated (e.g., GMSK for GSM) and equalized prior to channel decoding; these functions may be performed by a mix of ASIC and DSP. Channel decoding includes deinterleaving and error correction (e.g., Viterbi decoder), with the final step of voice decoding producing a raw stream of information bits.

Functions reciprocal to those used in the receiver are performed for the analog transmitter chain. For convenience and to reduce parts count the receiver and transmitter subsystems will share common intermediate frequencies and local oscillators.

The single carrier case is expanded to a multicarrier system, shown in Figure 2.2. The architecture is reasonably easy to scale with extra capacity being achieved by adding RF carrier transmit and receive chains. For narrow band systems (e.g., 30–200 kHz) this has the disadvantage that many chains are required to achieve moderate to large capacity. More chains push up the

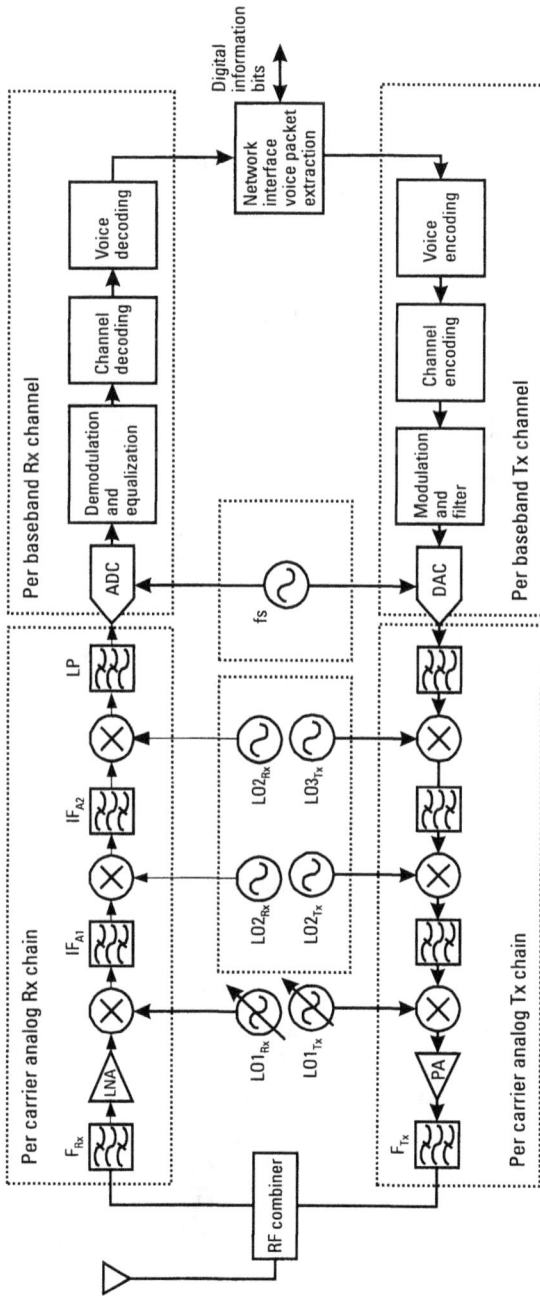

Figure 2.1 A 1990s hybrid analog and digital radio.

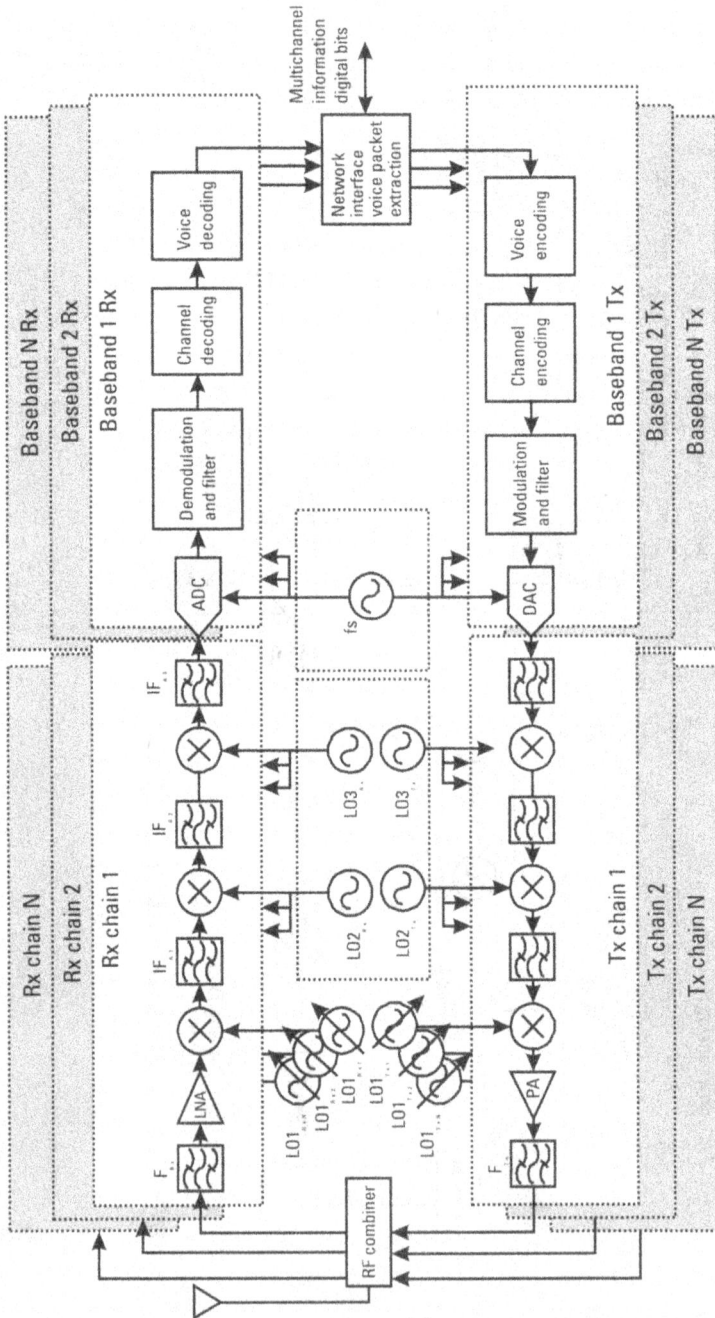

Figure 2.2 Multicarrier 1990s digital radio.

parts counts, power consumption, and maintenance requirements. In most cases the transmitter and receiver chains are combined into a single line replaceable unit (LRU) and referred to as a TRX.

Typically the baseband digital processing section consists of a signal processing card and is designed to handle a carrier's complement of transmit and receive processing. As with the analog front end of the system, this digital back end is also easily expandable by adding more baseband cards. The hardware complexity of high-capacity, multicarrier radios using this type of architecture can be significant. This complexity may not be a disadvantage when the protocols used for communication stay fixed over a reasonable period of time. A problem can occur when protocol changes are more rapid, particularly if an insufficient period has been allowed for amortization of the expenditure on the system equipment. This is one of the key problems that SDR promises to solve for 3G and beyond.

2.3 Basic Software Defined Radio Block Diagram

The closest we have ventured toward software radio is the hybrid case presented in the previous section, where, at best, narrowband modulation and demodulation functions are performed by DSP software at baseband. The best case includes the narrowband standards of GSM and IS 136 but would not include IS-95 CDMA. While it was technically feasible during 2G to use DSP software to implement the baseband parts of IS-95 CDMA, the availability of ASIC chip sets during the 1990s precluded this as a popular design solution due to cost.

Although the baseband functions of a 2G radio may have been performed by DSPs in software, it is likely that performing all the equivalent baseband functions for 3G with current generation DSPs will not be cost effective. The main reason for this is the fact that radio modulation techniques have increased in complexity at a faster rate than the processing capability of DSPs. Evidence of this is a blurring of traditional silicon architectures where DSPs are now including on-chip coprocessors (e.g., Viterbi and Turbo) in an effort to boost their application-specific performance.

With an ideal SDR architecture all the radio's functions from the antenna to the information interface are performed and programmed by a single high-level software language using generic computing and signal processing hardware. The basic SDR with a wideband RF front end, as shown in Figure 2.3, goes a significant way toward meeting the ideal goal. This

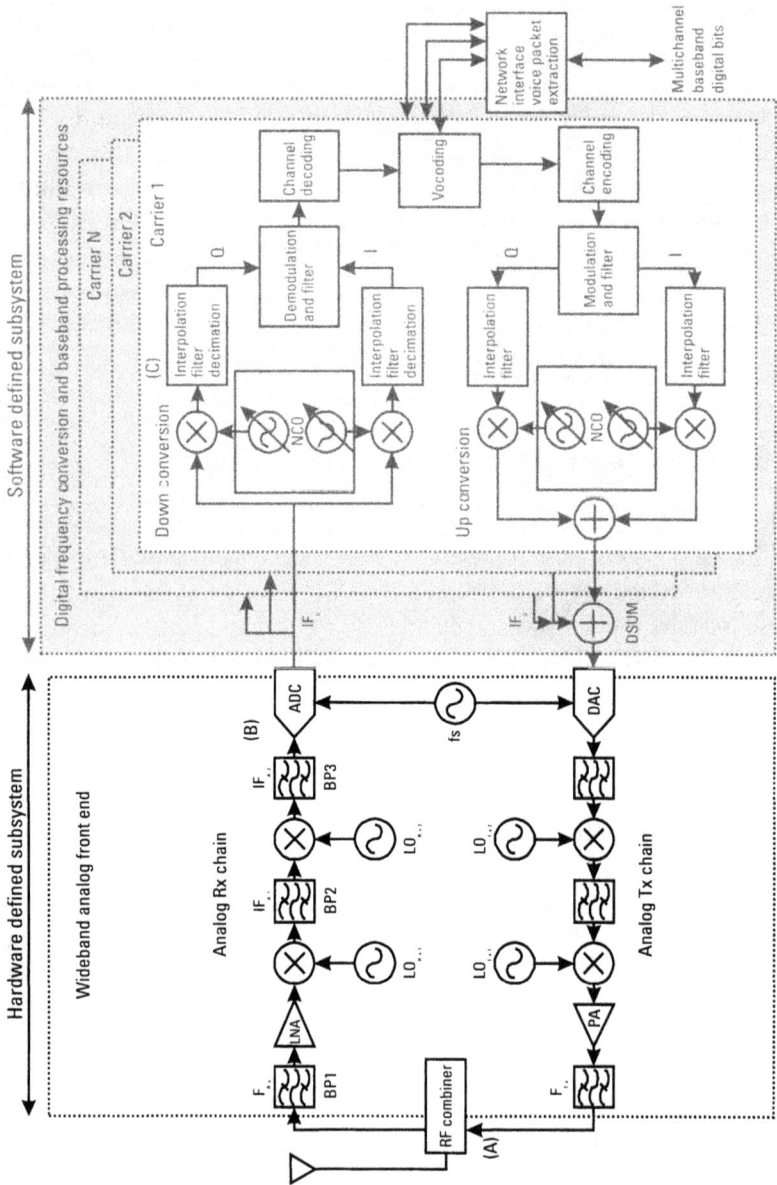

Figure 2.3 Basic SDR architecture with wideband RF front end.

functional block diagram is biased toward a multicarrier base station; however, the principles and major functions are the same for a single carrier terminal device.

The architecture in Figure 2.3 is divided into a hardware defined subsystem and a software defined subsystem. The hardware subsystem details some lower-level physical components (PA, LNA, ADC, and so on), whereas the software subsystem is purely functional and contains no indication of physical devices or lower-level partitions.

The hardware defined wideband analog front end is a complete subsystem. Its major difference from the front-end design of the hybrid radio shown in Figures 2.1 and 2.2 is its wideband capability, designed to replace many narrowband analog receive or transmit frequency conversion chains. Instead of shifting individual carriers to baseband (and filtering/rejecting the rest), the wideband front end converts or shifts an entire segment of spectrum to a suitable intermediate frequency—IF_D, "the digital IF"—prior to digitization. This process is graphically illustrated in Figure 2.4 by showing both positive and negative frequencies.

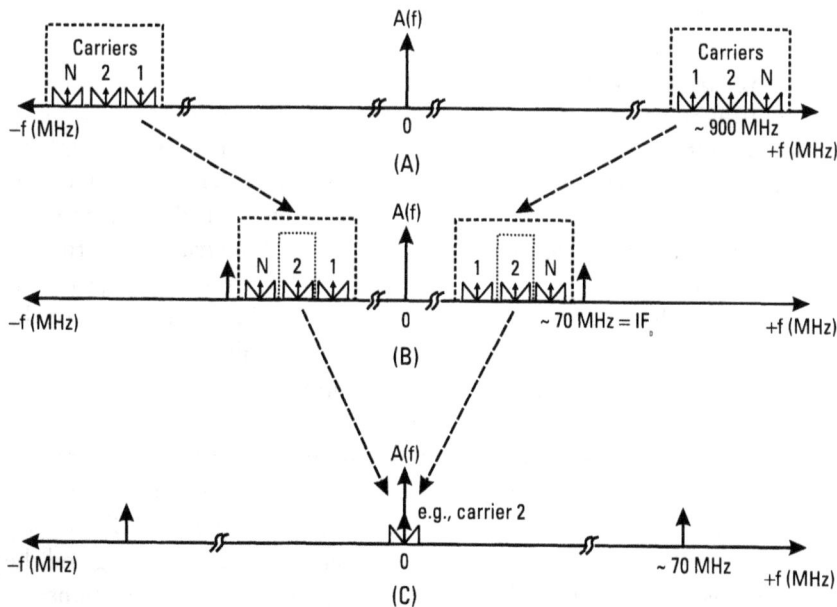

Figure 2.4 Wideband downconversion: (a) RF band, (b) Digital IF=IF subscript D, and (c) baseband.

Reference point (A) in Figure 2.4 shows a typical N channel group of carriers occupying a segment of the 900-MHz band, as would be the case for GSM. For the receive case the wideband analog front end shifts the entire N carriers to the digital intermediate frequency, or IF_D, as shown in (B). The choice of digital IF is dependent upon several factors, including the performance, cost, and availability of the antialiasing bandpass filters, BP3. A popular choice for IF_D is 70 MHz due to the COTS availability of satellite/microwave frequency converters (e.g., Vertex RSI [2] and SAW filters to perform the BP3 function). Finally, (C) in Figure 2.4 shows the spectrum of the required carrier (e.g., Carrier 2) shifted to baseband by software prior to demodulation; all other carriers have been completely suppressed. Points (A), (B), and (C) are referenced in Figure 2.3.

The software defined digital subsystem consists of digital frequency conversion and baseband processing. For receive the wideband multicarrier signal from the analog to digital converter is distributed to each of the frequency downconverters for processing. The downconverters isolate the required carrier and convert it to orthogonal I and Q baseband signals. The creation of in-phase (I) and quadrature-phase (Q) signals assumes that this SDR is implementing a digital modulation scheme. Phase quadrature is used either to create constant amplitude signals (e.g., GMSK), which reduce peak to average power ratios or increase bandwidth efficiency (e.g., QPSK) by transmitting more bits per hertz. The frequency conversion function is achieved digitally by numerically controlled oscillators (NCOs), multipliers, interpolators, decimation, and filtering. The software defined subsystem is data driven, because the frequency and timing reference is derived from the samples produced by the analog to digital and digital to analog converters.

Interpolation is an important feature in digital communications. This is because the digital sampling rate needs to be an integer multiple of the modulation symbol rate so that receiver and transmitter can achieve and maintain synchronization. If the SDR is to support multiple carriers and/or multiple symbol rates with fixed local oscillators, it is inevitable that the sampling rate will not coincide with an integer multiple of the symbol rate. For the case of fractional rates (e.g., 3.6 times), the oversampling rate needs to be lifted to the next integer multiple (e.g., 4 times) by interpolation. The choice of 2X, 3X, or 4X oversampling is a tradeoff between ease of filtering and the signal processing capability of the platform to perform more calculations.

Transmit-side functions are the reciprocal of the receive functions where baseband information is voice coded, channel coded, modulated, digital frequency upconverted, digital-to-analog converted, analog upconverted, and power amplified before transmission. The DSUM function in Figure 2.3 sums all the digital carriers to produce the multicarrier wideband signal that is processed by a single analog transmit chain and a single multicarrier power amplifier. Care must be taken during the design of the digital summing stage to avoid numerical overflow. The peak to average ratio of the multicarrier signal will depend on the system's air interface modulation and multiple access method; see Chapter 3 for more detail. The designer can either provide more bits to cope with the growth caused by summing (e.g., two 16-bit additions produce a 17-bit result) or scale the inputs to the summer to avoid an overflow.

The radio architectures presented in this chapter have not explicitly included any diversity capability. For most base stations the receive function in the subsystem will be duplicated to provide for receive space diversity. Receive diversity systems have a second receive chain connected to a spatially separated antenna. The digital processing section of the SDR then chooses the best received signal in order to reduce the effects of destructive interference caused by multipath fading.

2.4 System-Level Functional Partitioning

The SDR architecture shown in Figure 2.3 allocates analog hardware defined processing between the antenna and digital conversion (ADC and DAC) functions; while digital processing by software is allocated between digital conversion and the network interface. The more detailed RF design coverage provided in Chapter 3 can be used to assist with the subsystem level of partitioning for the analog part of the system, and Chapter 4 should be referenced when choosing the design parameters for the digital conversion function.

For the basic SDR we can further divide the digital subsystem into two major functions: digital frequency conversion and baseband processing. Let us consider some options for digital frequency conversion.

2.4.1 Digital Frequency Conversion Partitioning

The choice of sampling resolution (i.e., number of bits), sampling frequency, and digital IF bandwidth will drive the digital frequency conversion design solution. Mobile wireless communications systems generally do not benefit from sampling resolutions greater than 16 bits. This is because the hostility of the mobile RF environment (noise, multipath fading, and Doppler shift) prevents the use of sophisticated modulation schemes (e.g., 1,024 QAM),

Table 2.1
Data Processing Requirements

Standard	Oversampling Rate	Baseband Data Rate
IS 136	4X	0.78 MB/sec
GSM	4X	4.32 MB/sec
IS-95	4X	19.7 MB/sec
UMTS	4X	61.4 MB/sec

which, given benign channel conditions, may benefit from a larger number of bits of resolution. It is no mistake that most DSP processors perform fixed-point arithmetic and use a data width of 16 bits or 2 bytes. This is because most signal processing functions can be performed within the dynamic range that 16 bits provides.

Assuming that the digital IF is in the region of 70 MHz, the digital frequency conversion stage (up and down) must accept 140 MB/sec at the digital conversion interface; after processing the conversion stage must produce the baseband data bandwidths listed in Table 2.1. The UMTS figure of 61.4 MB/sec is calculated by multiplying the sample size in bytes (2) by the oversampling rate (4) by the chiprate (3.84×10^6) by the number of quadrature channels (I + Q = 2).

2.4.1.1 Partitioning Digital Frequency Conversion to DSPs

The next step in the design process is to choose a technology to host the frequency conversion function. The ideal SDR should comply with the software requirement—that is, the SDR should be completely reconfigurable by software. There are several candidate frequency conversion hosts, each fulfilling the software requirement to a varying degree. We will consider the DSP and programmable digital frequency converter (DUC and DDC) in this chapter; other potential candidates include the generic microprocessor and FPGA.

The software requirement can be met by partitioning the digital frequency conversion stage onto signal processing devices capable of being programmed in either assembler or C. Table 2.2 estimates the computational load that needs to be supported by these devices.

The MIPS count assumes that a multiply and accumulate (MAC) operation can be achieved in a single clock cycle, and, therefore, MIPS and MMACS are interchangeable in this context. The numerically controlled

Table 2.2
Computational Load

Function	Calculation	Computational Load
I Channel Freq Conversion	70×10^6 multiply and accumulate/sec	70 MIPS
I Channel NCO	$70 \times 10^6 \times 5$ instructions per frequency conversion	350 MIPS
Q Channel Freq Conversion	70×10^6 multiply and accumulate/sec	70 MIPS
Q Channel NCO	$70 \times 10^6 \times 5$ instructions per frequency conversion	350 MIPS
I Channel Interpolation	$70 \times 10^6 \times 5$ instructions	350 MIPS
Q Channel Interpolation	$70 \times 10^6 \times 5$ instructions	350 MIPS
I Channel Filtering (Decimation + FIR)	$70 \times 10^6 \times 20$ instructions	1,400 MIPS
I Channel Filtering (Decimation + FIR)	$70 \times 10^6 \times 20$ instructions	1,400 MIPS
	TOTAL	4,340 MIPS

oscillators may be implemented using lookup tables, and simple linear inter-polation could be used to reduce the number of required clock cycles. The finite impulse response (FIR) filtering load will depend on the required transfer function parameters (e.g., pass-band ripple and stop-band rejection). Table 2.2 indicates that all calculations are performed at a 70-MHz rate; however, some processing functions could be conducted at reduced rates by careful use of decimation to remove useless data.

Taking the computational load calculation result at face value, it seems that if the receive digital frequency conversion function were partitioned to DSP, it would consume at least one high-end DSP (e.g., TIC6X or Tiger-Sharc). If the DSP were clocked at 600 MHz and implemented instruction pipelining where four instructions were possible on every clock cycle, the DSP could yield 2,400 MIPS/MMACS; detailed DSP performance for a range of devices is provided in Chapter 6.

Obviously there is not enough processing power in a single DSP to handle the required 4,340 MIPS. Therefore, since all the functions could not be safely partitioned to a single high-power DSP, the designer would be forced to use an array of DSPs connected by an appropriate data transfer mechanism—for example, dual-port RAMs (DPRs). Good design practice allows a 50% spare capacity, which implies that this design case would require four 2,400 MMAC DSPs.

Although it has been suggested that the C language could be used to program the DSP, it is almost certain that the software developer would have to resort to hand-crafted assembler to have any chance of achieving a successful implementation and maintaining a safe spare capacity margin.

The major advantage of using a DSP is that the software requirement is fully complied with and complete flexibility is achieved. The disadvantage is that the designer is trading off flexibility for increased cost, power, and space consumption when compared with the solution of using application-specific devices such as programmable digital frequency up and downconverters (see Chapter 5 for more DDC and DUC detail).

2.4.1.2 Partitioning Digital Frequency Conversion to Application-Specific Devices

The DSP solution for digital frequency conversion is likely to consume four high-end DSPs and three DPRs per receive and transmit functional chain. In hardware terms this would likely cost several hundred dollars, 8W to 16W of power, and approximately 20 cm^2 of board real estate. One can see the difficulty in justifying the design, given that DDC and DUC devices can perform at least eight times the digital frequency conversion processing of a single DSP device. A single DDC device can typically process four channels of 70-MHz sampled data; using Table 2.2 this corresponds to 17,360 MIPS of digital frequency conversion capacity. The DDC will be roughly the same cost and size as a DSP and consume a similar quantity of power.

The mostly static frequency conversion functions listed in Table 2.2 require a small set of instructions to be performed in tight loops at very high data rates. DSPs simply do not contain enough parallel multiply and accumulate hardware to compete effectively with DUCs and DDCs in the mobile cellular communications domain. Although these application-specific devices are programmable within limited bounds, they are designed to meet the performance requirements of SDRs implementing the popular mobile air interfaces (e.g., UMTS, CDMA2000, GSM, IS 136, IS-95, EDGE, SCDMA, and so on).

For cellular mobile wireless systems operating above the 20–100 MHz range, DUCs and DDC are currently the most efficient bridge between digital intermediate frequencies and baseband processing. These devices are also sufficiently programmable to fit within the SDR definition.

2.4.1.3 Baseband Signal Processing Partitioning

With the arrival of a CDMA dominated 3G, baseband signal processing is now commonly divided into chip rate and symbol rate processing functions.

Receive chip rate processing includes demodulation and the CDMA despreading operations; the result is a raw bit stream including the channel coding overhead. Symbol rate receive processing performs all the channel decoding functions (e.g., deinterleaving, Viterbi decoding, and so on) to produce an information bit stream.

The issues to be considered during this aspect of partitioning will be largely driven by the air interfaces that must be supported by the baseband processing resource platform. From the receive processing perspective, Table 2.1 illustrates the huge range of input baseband data rates for the various air interface standards. After chip and symbol rate processing, the output information data rate can vary from 13 Kbps for compressed and vocoded voice to 2 Mbps for exceptional quality 3G data connections.

Candidate resource platform technologies include microprocessors, DSP, reconfigurable processors, and FPGA. In most cases the microprocessor will either have insufficient bandwidth and performance for the required tasks or it will consume too much power and space to be cost effective. FPGAs and reconfigurable processors (e.g., Chameleon RCP) can effectively meet the processing bandwidth requirements; however, they are more difficult to program than DSPs. A detailed overview of signal processing technologies is provided in Chapter 6.

Non-CDMA 2G baseband processing is entirely possible within DSP alone because of the reduced data bandwidth requirement. However, DSP does not have the processing performance for chip-rate 3G processing and is best suited to handling the symbol-rate functions. The complete baseband partition for a UMTS 3G system could use DSPs for symbol rate and FPGAs or RCPs for chip rate.

2.5 A COTS Implementation of the Basic Architecture

There will be software defined radio developers who either do not plan to develop their own digital processing platform or wish to start implementation of their system as quickly as possible. For these situations it is possible that a COTS solution will suit the system requirements. The recent interest in software defined radio has been picked up by several third-party board vendors. Evidence of this is found in direct software radio references starting to appear on manufacturers' Web sites (e.g., Software Radio Central [3] and software radio [4]).

Pentek is one of the more SDR biased suppliers, with the company making available a wireless cellular development system for software radio

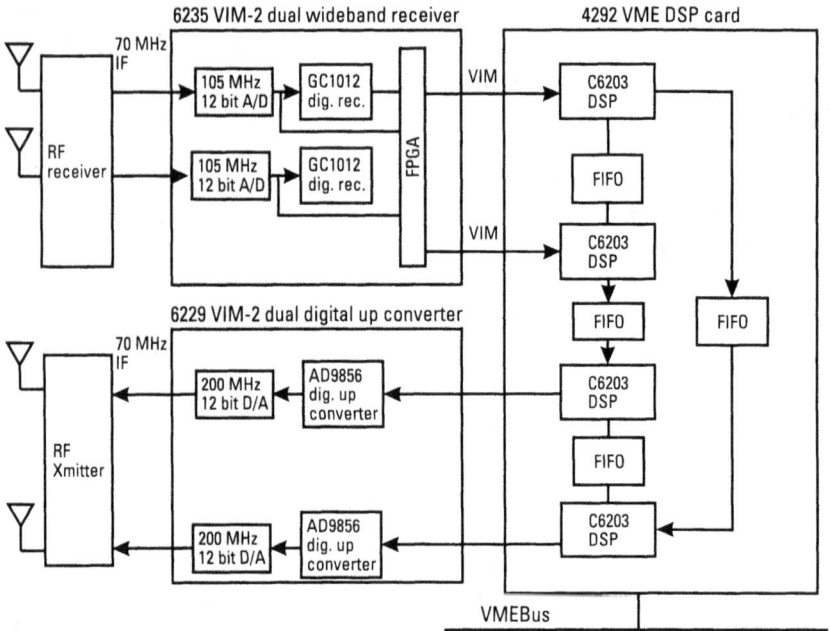

Figure 2.5 Wireless cellular development system.

developers. This collection of hardware has the capability to perform the digital frequency conversion and baseband processing segment of the basic SDR architecture with a wideband RF front end, as depicted in Figure 2.5.

The wireless cellular development system architecture uses the VME form factor, an open commercial bus standard managed since 1984 by the VMEBus International Trade Association (VITA) [5]. VME started life in 1979 when it was developed by Motorola for its 68000 microprocessor. Since then it has developed into a popular high-speed bus standard supported by many manufacturers and used in hundreds of products. The Pentek system uses the VME64 variant, which supports 80 MB/sec transfers; the standard includes the even faster VME320, which can deliver transfers between 320 and 500 MB/sec. VME is a popular COTS choice, because it has been around for many years and there are now many suppliers of competing products, including DSP boards, single board computers (SBC), and network interfaces.

The heart of the development system is the 4292 Quad TMS320C6203 DSP processor VME board; see Figure 2.6 for a detailed schematic. Figure 2.5 shows a basic block diagram of the 4292 with four onboard C6203 DSPs and four interconnecting first-in-first-out (FIFO) buffers. The board has major external interfaces to four velocity interface mezzanine (VIM) sites plus VMEBus and RACEWay. The VIM interface suits software radio because of its fast 300 MB/sec transfer rate. RACEWay is an ANSI/VITA industry standard with 267 MB/sec potential throughput; the interface can be used to interconnect with other DSP boards in a larger system.

Digital frequency conversion and translation from digital IF to baseband is performed by the dual wideband receiver and dual digital upconverter, which plug directly into the VIM interfaces as daughter modules to the DSP card. This solution for frequency conversion using dedicated digital up and downconverters reinforces the conclusion in Section 2.4, where it was considered more effective to use these devices than assigning the function to DSPs.

The wideband receiver contains two 105-MHz, 12-bit analog to digital converters (ADCs), two wideband digital downconverters (DDCs), and a 500,000 to 3 million gate FPGA. Using a 70-MHz analog intermediate frequency, the ADCs can be designed for lowpass or bandpass sampling depending on the required system bandwidth (see Chapter 4 for more ADC detail). The DDCs can deliver the FPGA with bandwidths in the range of 1.25 to 40 MHz. For narrower band signals (e.g., 30 kHz IS 136) the FPGA can be programmed to perform additional filtering to reduce the signal processing burden on the DSPs. The FPGA is also available to perform the important interpolation function.

The DUC module contains two 200-MHz, 12-bit DACs and two 200-MHz interpolating digital upconverters.

This complete arrangement provides enough digital signal processing power to support major software radio functionality in a single VME slot. In most cases the actual system would occupy two VME slots with the second slot populated with a single board computer (SBC). With the addition of an analog wideband front end (RF Receiver and RF Xmitter in Figure 2.5), the developer has access to a complete COTS platform capable of SDR functionality for the implementation of diversity receivers and transmitters.

A detailed block diagram of the 4292 DSP board is provided in Figure 2.6, while Figure 2.7 illustrates the VME form factor of a similar 4290 DSP board complete with VIM expansion modules.

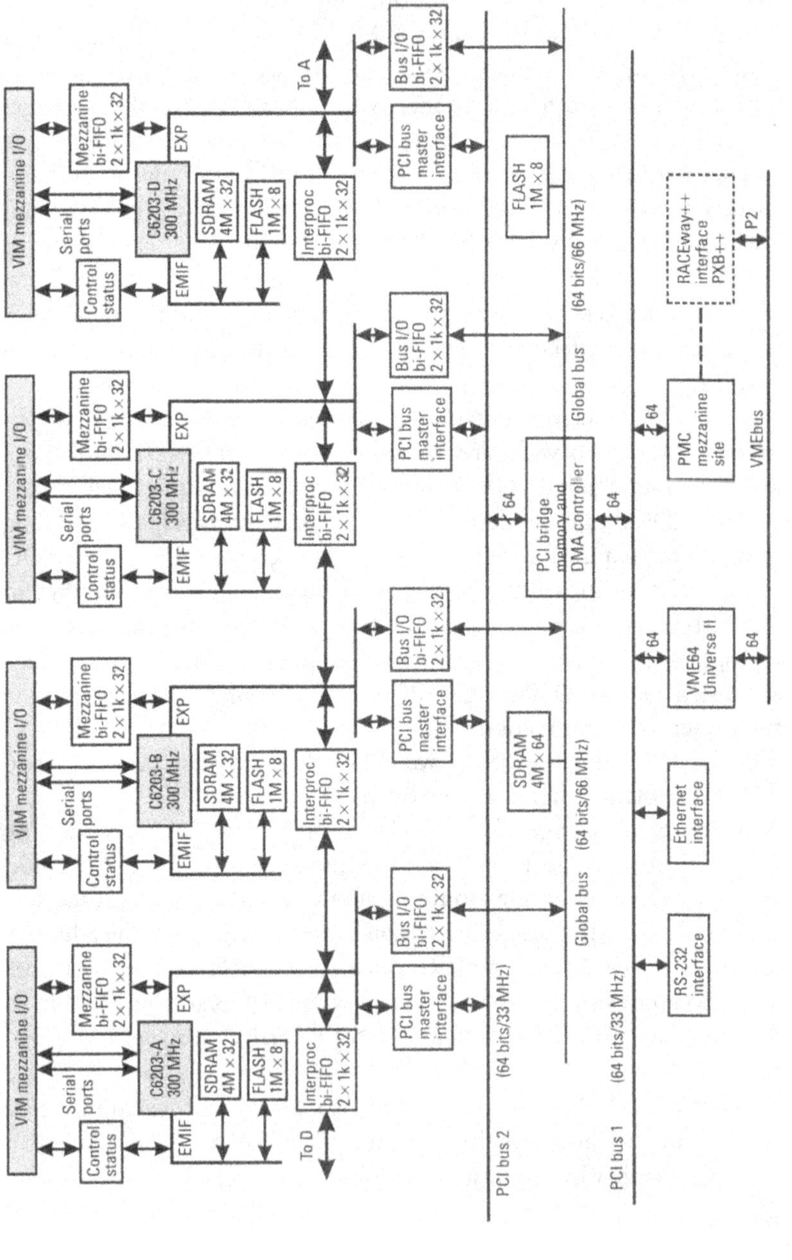

Figure 2.6 4292 DSP board block diagram. (*Source:* Pentek, 2001. Reprinted with permission.)

Figure 2.7   Pentek DSP board. (*Source:* Pentek, 2001. Reprinted with permission.)

2.6 Conclusion

This chapter has covered the progression of radio architectures from hardware defined analog systems through the hybrid analog and digital solution and concluded with a COTS implementation of a software defined radio. The basic SDR architecture can suit the needs of the 3G designer and more application-specific detail is provided on this topic in Chapter 8. We have focused on the underlying hardware platform capability in this chapter with little regard for the software that needs to run on it; this covers the bottom-up design approach. Please see Chapters 7 and 8, where the top-down approach is followed through and software is considered.

References

[1]   Tsurumi, H., and Y. Suzuki, "Broadband RF Stage Architecture for Software Defined Radio in Handheld Terminal Applications," *IEEE Communications Magazine*, February 1999.

[2]   http://www.rsicom.com.

[3]   http://www.pentek.com.

[4]   http://www.transtech-dsp.com/apps/radio.htm.

[5]   http://www.vita.com.

3

RF System Design

In this chapter we cover the RF aspects that must be considered during system design and concentrate on the impacts upon the analog and digital parts of a software defined radio. Radio propagation and real-world environmental effects play a big part during the design of a wireless system. The frequency of operation, power levels, physical terrain, interference, and other effects require a systematic approach during the design process.

3.1 Introduction

RF engineering is sometimes called the black art. However, the huge and growing demand for wireless systems is requiring more engineers from every discipline to understand the topics covered in this chapter. It is important to understand that analog electronics add noise and distortion to the information bandwidth. Many of the functions performed by digital electronics and software are designed to limit the effects of the analog processing (e.g., digital predistortion to correct for power amplifier nonlinearity). Therefore, the analog part of an SDR must have an achievable performance budget and be designed in concert with the digital subsystem.

The emissions caused by radio mean that SDR equipment must comply with the strict electromagnetic interference (EMI) and electromagnetic compatibility (EMC) requirements set by standards bodies (e.g., 3GPP and 3GPP2) and regional approval authorities (e.g., FCC). The SDR must be designed and built to meet the required type approval tests before the product can be operated in the target country.

The cellular network design, radio link budget, and analog/digital system partition in the SDR are all factors that have implications for the signal processing hardware and software resources. Once design decisions are made in the digital partition, they should be flowed back to the analog/RF partition to ensure that the system design and functional allocation are optimal. The conclusion of the design flow is an SDR that meets its functional and performance requirements and is cost effective to maintain and operate over its lifetime.

3.2 Worldwide Frequency Band Plans

The history of band planning for 3G dates back to the International Telecommunication Union's (ITU) World Administrative Radio Conference (WARC, 1992) held in Malaga, Torremolinos, Spain, in 1992. The conference allocated 1,885 to 2,025 MHz and 2,110 to 2,200 MHz as core IMT 2000 (3G) frequency bands on a worldwide basis; this is reiterated in the ITU's IMT 2000 Recommendations 1034 [1], 687 [2], and 1035 [3].

WARC 2000 identified a number of extension bands to add additional spectrum from 2005, (i.e., 806 to 960 MHz, 1,710 to 1,885 MHz, and 2,500 to 2,690 MHz). Because of the history of band planning, or more because of a lack of it, each country will implement different parts of the IMT 2000 bands at different times. The 2G uses another set of bands (e.g., GSM 400, GSM 900, GSM 1800, and GSM 1900) so users who switch back and forth between 2G and 3G can conceivably operate anywhere between 450.4 MHz for GSM 400 to 2,690 MHz for 3G. This large frequency spread is one reason why the ideal SDR for 3G is currently not a reality.

However, the need to support so many frequency standards makes SDR an even more appealing technology than it may have been if the world had been successful at organizing a single spectrum allocation per mobile cellular generation.

3.3 Noise and Channel Capacity

One of the most important parameters to be considered during the design of any communications systems is noise. In analog and RF systems, thermal noise is generated when electrons pass through conductors. Noise is also contributed by human-generated and naturally occurring events. Most importantly for CDMA systems, additional noise is generated by other users who are simultaneously sharing bandwidth by sharing power.

The major difficulty for cascaded analog systems is the additive nature of noise. Careful attention must be paid during the allocation of gain and noise figure to each processing block in a cascaded RF system. The limiting performance factor for digital systems is quantization noise as generated during the analog to digital conversion process. Quantization noise is propagated through a digital processing system, but it is not added to because the conversion process only happens once. Poorly designed digital systems can, however, suffer from overflow or rounding errors that can be translated into noise.

The importance of noise in a communications system is illustrated by the Shannon-Hartley theorem, as expressed by (3.1). The channel capacity, C, is measured in bits per second, and B is the channel bandwidth measured in hertz. S and N are the average signal power and noise power, respectively. Given a fixed signal power, S, the higher the noise level, N, the less information, C, that can be transmitted through a channel.

$$C = B \log_2 \left(1 + \frac{S}{N} \right) \qquad (3.1)$$

Thermal noise is defined by (3.2), where N is the noise power in watts, k is Boltzmann's constant (1.381×10^{-23} J/molecule K or W/Hz/K), T is the noise temperature in Kelvin, and B is the bandwidth of the channel measured in hertz.

$$N = k\,T\,B \qquad (3.2)$$

When working in decibels and referencing to milliwatts, k is equal to -198.6 dBm/Hz/K. For land-based radio systems, T will be in the vicinity of 290K and kT can be simplified to -174 dBm/Hz. For a UMTS channel bandwidth B of 5 MHz, thermal noise, N, will equal -107 dBm, and for CDMA-2000 1X, where $B = 1.23$ MHz, thermal noise, N, will equal -113.1 dBm.

3.4   Link Budget

In Figure 3.1 we show a typical radio system consisting of a transmitter and a receiver separated by some physical distance, d, measured in meters. The

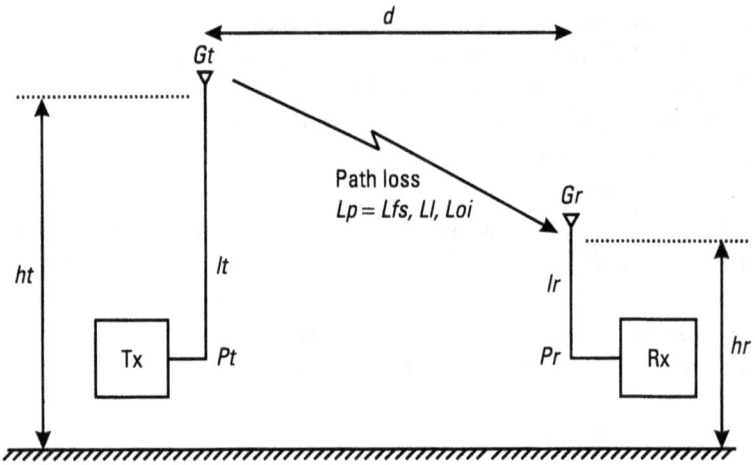

Figure 3.1 Simple system propagation loss block diagram.

base station transmitter is connected to its antenna with gain, *Gt*, by a transmission line with loss *lt*. The mobile receiver is connected to its antenna with
gain, *Gr*, by a transmission line with loss *lr*.

For most cases in a mobile radio system the base station antenna height
(*ht* in meters) will be higher than the mobile antenna height (*hr* in meters).
The propagation loss, *Lp*, will be a function of the transmission frequency *f*,
the distance *d*, and the terrain in-between the transmitter and receiver.

3.4.1 Free Space Loss

Assuming the receive antenna is separated from the transmit antenna by a
distance greater than the near field distance (i.e., greater than ten wave-
lengths), the free space loss, *Lfs*, can be calculated by (3.3).

$$Lfs(d) = -10\log\left[\left(\frac{\lambda}{4\pi d}\right)^2\right] \tag{3.3}$$

Free space loss is proportional to the distance squared and inversely
proportional to the wavelength squared. Knowing that the speed of light *c* =
fλ, and converting loss to decibels, (3.3) can be expressed as (3.4), where frequency *f* is measured in hertz.

$$Lfs(d) = -147.6 + 20\log(d) + 20\log(f) \tag{3.4}$$

Given a fixed frequency, free space path loss will increase at 20 dB per decade with increasing distance. The path loss for a typical 3G frequency of 1.9 GHz will be 78 dB at 100m and 98 dB at 1 km.

3.4.2 Practical Loss Models

The free space model is a best-case reference and possibly only used in mobile communications where a line of sight path exists over water. Paths that are not line of sight and include obstructions such as buildings, vegetation, and varying terrain require different and generally more complicated models to accurately predict path loss.

One such model is the general form of the model developed by Lee [4] and summarized by Yang [5]. This model assumes a suburban mobile radio environment and takes into account the effective gain introduced by differing heights of transmit and receive antennas. Equation (3.5) is used to calculate the path loss, *Ll*, in decibels.

$$Ll(d) = -159.4 + 20\log(f) + 38.4\log(d)$$
$$-20\log(ht) - 10\log(hr) \tag{3.5}$$

The Lee model indicates that path loss is more dependent on distance than the free space model. Assuming a frequency of 1.9 GHz, a transmit antenna height of 12m, and a receive antenna height of 2m, the path loss will be approximately the same as produced by the free space model: 79.6 dB at 100m. However this will increase by 38.4 dB per decade resulting in a larger loss of 118 dB at 1 km.

3.4.3 IMT 2000 Path Loss Models

As part of the definition of compatible IMT 2000 3G mobile technologies, the ITU issued Recommendations 687 [2] and 1225 [6]. In 1998 Recommendation 1225 provided evaluation criteria for the then-candidate "Radio Telecommunications Technologies" against the ITU's 3G requirements. Submissions in response to the ITU were made by several organizations, including the TIA, which submitted a response based on CDMA2000 [7], and

ETSI with its UMTS [8] submission. The successful IMT 2000 technologies have since been selected and include both UMTS and CDMA2000.

The ITU documents and IMT 2000 submissions contain a great deal of useful 3G design information relevant to RF systems. Appendix 1 of Annex 2 [7] specifies the following propagation models for IMT 2000 and 3G:

- Path loss model for indoor office test environment;
- Path loss model for outdoor to indoor and pedestrian test environment;
- Path loss model for vehicular test environment.

3.4.3.1 Outdoor to Indoor and Pedestrian Model

The total propagation loss, Loi, for the outdoor to indoor and pedestrian model is described by (3.6). The loss is the sum of the free space loss, Lfs; the diffraction loss from rooftop to the street, $Lrts$; and the reduction due to multiple screen diffraction past rows of buildings, $Lmsd$.

$$Loi(d) = Lfs(d) + Lrts(x, \Delta hm, d) + Lmsd(bs, d) \qquad (3.6)$$

The diffraction loss, Lts, is calculated by (3.7), where Δhm is the difference between the mean building height and the mobile antenna height and x is the horizontal distance between the mobile and the diffracting edges.

$$Lrts(x, \Delta hm, d) =$$

$$-10 \log \left[\frac{\lambda}{2\pi^2 r} \left(\frac{1}{a\tan\left(\frac{|\Delta hm|}{x}\right)} - \frac{1}{2\pi + a\tan\left(\frac{|\Delta hm|}{x}\right)} \right) \right]^2 \qquad (3.7)$$

where:

$$r = \sqrt{(\Delta hm)^2 + x^2}$$

and the multiple screen diffracting loss, *Lmsd*, is calculated by (3.8), where *bs* is the average separation between rows of buildings.

$$Lmsd\left(bs,d\right) = -10\log\left[\left(\frac{bs}{d}\right)^2\right]$$ (3.8)

The outdoor to indoor and pedestrian model are the most dependent on distance of all the models discussed. Assuming a frequency of 1.9 GHz, x = 5m, Δ *hm* = 10m, and *bs* = 20m, the path loss at a distance of 100m will be 125.7 dB and increase at 40 dB per decade to 165.7 dB at d equal to 1 km. Figure 3.2 plots loss for the three models presented.

The link budget for the simple system shown in Figure 3.1 can be calculated using (3.7) where power is measured in decibels (e.g., decibels per meter) and antenna gain is measured in decibels relative to an isotropic gain of one. The received power, *Pr*, at the input to the receiver is equal to the transmitted power *Pt*, plus the antenna gains *Gt* and *Gr*, minus the cable losses *lt* and *lr*, minus the propagation loss *lp*. The cable loss, *lt* and *lr*, is con-

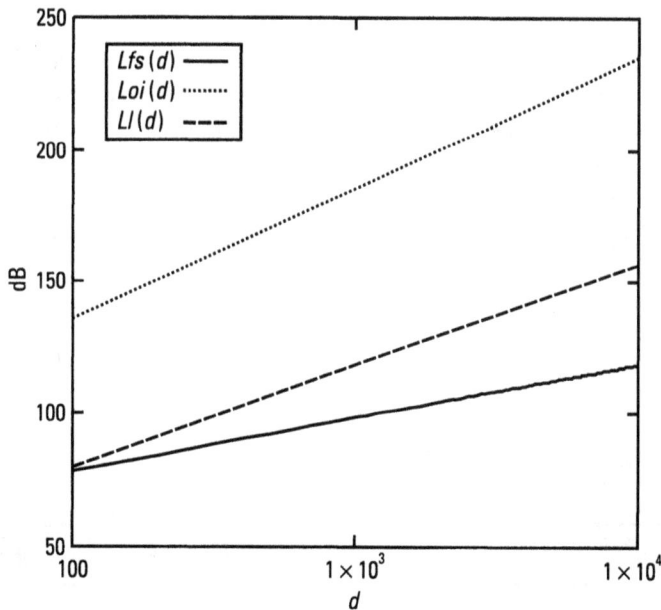

Figure 3.2 Propagation loss versus distance.

sidered to include the associated combiner loss for systems that use a single antenna for both transmit and receive, as is the case in a mobile phone.

$$Pr = Pt - lt + Gt - Lp + Gr - lr \qquad (3.9)$$

Using link budget figures from Section 8.1 [7] and assuming the IMT 2000 outdoor to indoor and pedestrian model, we will now calculate a typical received power, Pr, for the forward link. Given a transmit power of Pt of 30 dBm, Gt = 13 dBi, Gr = 0 dBi, lt = 2 dB, lr = 0 dB, and d = 100m, the received power is 30 dBm – 2 dB + 13 dB – 135.2 dB + 0 dB – 0 dB = – 94.2 dBm.

3.4.4   Detailed System Link Budget

To determine appropriate gains for the lower-level analog and RF functional blocks in the radio system we need to develop a more complete model. Figure 3.3 details the typical analog and RF functional blocks between the digital parts of the radio system (i.e., between A and D).

Figure 3.3 shows the output of the base station digital to analog converter at reference point A driving the transmit subsystem with power Pa. The output of the transmit subsystem at reference point B drives the transmit transmission line with power Pt. The input to the receive subsystem at reference point C receives power Pr, and the output of the receiver at reference point D drives the mobile's analog to digital converter with power Pd.

The transmit subsystem consists of two cascaded frequency upconverters followed by a power amplifier with gains of $Gtx1$, $Gtx2$, and $Gtx3$,

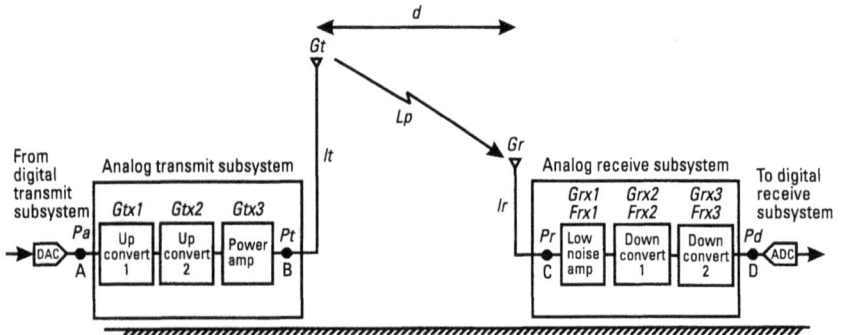

Figure 3.3   Expanded RF system block diagram.

respectively. The receive subsystem consists of a low-noise amplifier followed by two cascaded frequency downconversion stages with gains of $Grx1$, $Grx2$, and $Grx3$, respectively.

Equations (3.10) and (3.11) calculate the powers seen at reference points B and D, respectively.

$$Pt = Pa + Gtx1 + Gtx2 + Gtx3 \tag{3.10}$$

$$Pd = Pr + Grx1 + Grx2 + Grx3 \tag{3.11}$$

Assuming that the analog-to-digital converter power output, Pa, is 0 dBm, a required Pt of 30 dBm could be achieved by equal gain distribution with $Gtx1$, $Gtx2$, and $Gtx3$ all set to 10 dB.

Designing the receive subsystem cannot be made on the basis of gain selection alone. One must consider the cumulative noise effect, as introduced in Section 3.2. A useful performance parameter is noise figure. Consider you have a device with input signal-to-noise ratio, Si/Ni, and an output signal-to-noise ratio of So/No. The noise figure, F, of the device is simply the input signal-to-noise ratio divided by the output signal-to-noise ratio, as expressed by (3.12).

$$F = (Si \div Ni) \div (So \div No) \tag{3.12}$$

Practical devices such as amplifiers and mixers always exhibit a smaller signal to noise ratio at their output compared with signal to noise ratio at the input. Therefore, F will always be larger than one. It has been shown [9] that the total noise figure, F_t, for a cascaded system is calculated by (3.13), where F_1 and G_1 are the noise figure and gain, respectively, of the first element in the chain, and Fn and Gn are the noise figure and gain of the nth element of the cascaded chain.

$$F_t = F_1 + \frac{(F_{2-1})}{G_1} + \frac{(F_3 - 1)}{G_1 G_2} + \ldots + \frac{(Fn - 1)}{G_1 G_2 \ldots G_{n-1}} \tag{3.13}$$

Equation (3.13) demonstrates that noise figure is dominated by the gain and noise figure of the first element.

3.4.4.1   Transmit Subsystem Noise Figure

Quantization noise will be injected into the RF transmit subsystem by the digital to analog converter and each processing stage of the subsystem will add more noise. However, calculation of transmitter noise figure is not usually a concern for the cellular mobile RF system engineer. This is because the thermal noise contributions from the transmit subsystem will be attenuated by an amount equal to the propagation loss after arriving at the receive antenna. The ambient thermal noise seen by the receive antenna is usually much greater than the attenuated transmitter-borne noise and is the dominant noise contribution affecting signal detection performance and bit error rate. This analysis assumes that the transmit subsystem uses a similar number of processing stages to the receive subsystem and has a similar noise figure.

3.4.4.2   Receiver Subsystem Noise Figure

The UMTS and CDMA2000 ITU-RTT Candidate Submissions [7, 8] suggest a receiver noise figure of 5 dB for both uplink (mobile to base station) and downlink (base station to mobile). If we were to design the receiver subsystem for Figure 3.3 and use $lr = 0$ dB, the noise figure requirement would be set at 5 dB less an implementation margin (in the range of 1–2.5 dB). Using (3.12), we could choose the appropriate noise figure and gain for the low noise amplifier and the two frequency downconverters.

3.5   3G RF Performance Requirements

Performance specifications and test requirements for 3G terminals and base stations can be found in the standards for CDMA2000 [10, 11] and UMTS [12, 13]. The following paragraphs cover the key RF parameters included in these standards.

3.5.1   Receiver Requirements

Receivers are characterized by many more parameters than just gain and noise figure. The 3G specifications define these as follows.

3.5.1.1   Sensitivity

This is the minimum receiver input power measured at the antenna input connector for which the BER does not exceed a specified value. The UMTS base station receiver sensitivity specification is –121 dBm, given a data rate of 12.2 Kbps and a BER of less than 0.001.

3.5.1.2 Dynamic Range

This is the ability of the receiver to handle a surge of interference in the channel. The receiver is required to fulfill a specified BER requirement for a specified sensitivity degradation of the required signal in the presence of an interfering additive white Gaussian noise (AWGN) signal in the same channel.

A UMTS base station receiver must be able to maintain a BER of less than 0.001, while the required signal power is –91 dBm or greater, and the interfering signal power is –73 dBm.

3.5.1.3 Adjacent Channel Selectivity

Adjacent channel selectivity is a measure of the receiver's ability to receive a wanted signal on its assigned channel in the presence of an adjacent channel signal at a given frequency offset from the center frequency of the assigned channel.

For a BER of less than 0.001 and a data rate of 12.2 Kbps, a UMTS base station receiver with a wanted signal power of –115 dBm must be able to handle an interfering signal power of –52 dBm, given that the interferer is either 5 MHz above or below the wanted signal.

3.5.1.4 Blocking

Blocking is a measure of the receiver's ability to maintain performance for a wanted signal in the presence of an interferer. The interferer is located on any frequency other than the adjacent channel or those coinciding with the receiver's spurious responses. For example, in the 1,920–1,980-MHz band, a UMTS base station receiver's performance should not degrade, given a wanted signal power of –115 dBm and an interfering WCDMA signal power of –40 dBm at a 10 MHz offset from the wanted signal.

3.5.1.5 Intermodulation

The mixing of two interfering RF signals can produce second-, third-, and higher-order products in the band of the wanted channel. Intermodulation rejection is a measure of the capability of the receiver to receive a wanted signal on its assigned channel in the presence of two or more interfering signals that have a specific frequency relationship to the wanted signal.

Table 3.1
UMTS Intermodulation Specification

Type of Signal	Offset	Signal Level
Wanted signal	—	–116 dBm
CW interferer	10 MHz	–48 dBm
WCDMA interferer	20 MHz	–48 dBm

A UMTS base station receiver with a wanted signal of –116 dBm must be able to maintain performance (BER < 0.001) in the presence of interferers, as specified in Table 3.1.

3.5.2   3G Transmitter Requirements

The following paragraphs define and summarize the key parameters that characterize the performance of a 3G transmitter.

3.5.2.1   Occupied Bandwidth

The transmitter's occupied bandwidth is the span of frequency that contains a specified percentage of the mean emitted power. For a UMTS base station the occupied bandwidth must be less than 5 MHz and must contain 99.5% of the mean emitted power. In other words, only 0.5% of the mean power is allowed to be radiated outside the occupied bandwidth.

3.5.2.2   Out-of-Band Emissions

These emissions are the unwanted emissions occurring immediately outside the desired channel bandwidth. Out-of-band emissions exclude spurious emissions and result from the modulation process and nonlinearity in the transmitter. The out-of-band emission limit is specified in terms of a spectrum emission mask and adjacent channel leakage power ratio for the transmitter.

Part of the UMTS base station specification states that for a transmitter with power output of greater than 43 dBm, at an offset of 7.5 MHz from the carrier, a 1-MHz measurement filter shall record a power of less than –13 dBm.

3.5.2.3   Transmitter Spurious Emissions

Spurious emissions are emissions that are caused by unwanted transmitter effects, such as harmonic emissions, parasitic emissions, intermodulation products, and frequency conversion products, but exclude out-of-band emissions.

For a UMTS base station the requirement applies at frequencies that are more than 12.5 MHz under the first carrier frequency used and more than 12.5 MHz above the last carrier frequency used. Part of the requirement states that in the 150 kHz to 30 MHz region spurious emissions shall be less than −36 dBm in a 10-kHz measurement bandwidth.

3.5.2.4 Transmitter Intermodulation

Intermodulation performance is a measure of the capability of the transmitter to inhibit the generation of signals in its nonlinear elements caused by the presence of the wanted signal and an interfering signal reaching the transmitter via the antenna.

For a UMTS base station the intermodulation level is the power of the intermodulation products when a WCDMA modulated interference signal is injected into an antenna connector at a level of 30 dB lower than that of the wanted signal. The frequency of the interference signal can be 5 MHz, 10 MHz, or 15 MHz offset below the first or above the last carrier frequency used in the transmitter.

The transmit intermodulation level must not exceed the out-of-band emission or the spurious emission requirements.

3.6 Multicarrier Power Amplifiers

The 3G mobile terminals that support multiple air interfaces (e.g., GSM and UMTS) will provide challenges for the subwatt power amplifier designer. However, it is possibly the 3G SDR base station with its multicarrier high-power amplifier that poses the biggest design challenge and offers the biggest system improvement. A major benefit of multicarrier SDR is that an array of single carrier amplifiers and power combiners can be replaced by a single multicarrier unit that takes up less space, consumes less power, and costs less to own and operate over its lifetime.

3.6.1 Power Amplifier Linearizers

Modulation schemes that exhibit a constant envelope modulus (e.g., GMSK) with low peak to average power ratios require power amplifiers with minimal linearity specifications. This is because it is easier to operate power transistors over a smaller dynamic range and maintain linearity. From the link budget

Table 3.2
Peak to Average Power Ratios

Air Interface	Typical Signal Peak-to-Average Ratio
AMPS single carrier	0 dB
GSM single carrier	1.5 dB
TDMA single carrier	3.5 dB
IS-95 CDMA single carrier	10 dB
WCDMA/UMTS single carrier	8–9 dB
IS-95 CDMA multicarrier	10.5 dB
WCDMA/UMTS multicarrier	12.2 dB
EDGE multicarrier	9 dB

perspective, only the average power is considered in any calculations [refer to (3.1) and Shannon's information theorem].

QPSK has better spectral efficiency but also a higher peak to average power ratio; this is because the signal constellation can transition through zero and cause large changes in instantaneous amplitude. For a multicarrier system, adding two or more GMSK carriers at different frequencies destroys the constant modulus property, thereby increasing dynamic range and the need for better linearity.

Table 3.2 [14, 15] lists typical peak to average power ratios for several 1G, 2G, and 3G scenarios.

The 1G AMPS system uses FM; the transmitted waveform is absolutely constant in amplitude, which produces a 0-dB peak-to-average ratio. Simple GMSK would also produce a 0-dB ratio, but the GSM system uses coarse, grained, slow rate (relative to CDMA) power control, which adds about 1.5 dB to the peak to average power ratio. A multicarrier UMTS/WCDMA signal produces a 12.2-dB peak to average ratio compared with approximately 10.5 dB for a multicarrier CDMA2000 signal. The UMTS scenario exhibits higher peak to average values because of its wider band nature (i.e., 5 MHz) when compared with EDGE (200 kHz) and IS-95/CDMA2000 (1.25 MHz).

In general, 3G multicarrier amplifiers must have better linearity performance than 2G multicarrier amplifiers, and 2G and 3G amplifiers must have better linearity performance than 1G power amplifiers.

In the past, single channel power amplifier design was a relatively simple design task. Linearity requirements could be handled by increasing the dc biasing of the final power transistors. Classes A, AB, B, and C alter the transistor biasing, which changes the utilization of the power curve. Class C is adequate for some single carrier FM transmitters, no dc biasing is used and the power efficiency is relatively high (approximately 40–70%). With Class A amplifiers the dc bias is increased so that the output waveform only occupies the most linear portion of the transistor's power curve. Class A is often coupled with selective use of power back-off. This is equivalent to selecting transistors with much higher power rating than required in an attempt to gain more linearity over a larger amplitude range. Class A amplifiers consume significant power even when there is no input signal present and hence the efficiency is lower; with back-off the efficiency can be in the range of 10–40%. A disadvantage of back-off is that the technique is limited to lower powers and the cost of transistors is increased, sometimes to the point where the technique is not commercially viable.

Many complicated techniques for improving linearity have been developed over the years as modulation schemes have become more complex and multicarrier systems more prevalent. We will now discuss one popular technique for improving linearity.

3.6.1.1 Feed-Forward Linearization

The feed-forward technique is inherently wideband and is used in 1G and 2G multicarrier power amplifiers for AMPS, TDMA, and GSM. As shown in Figure 3.4, the design has two major signal paths. The input low-power

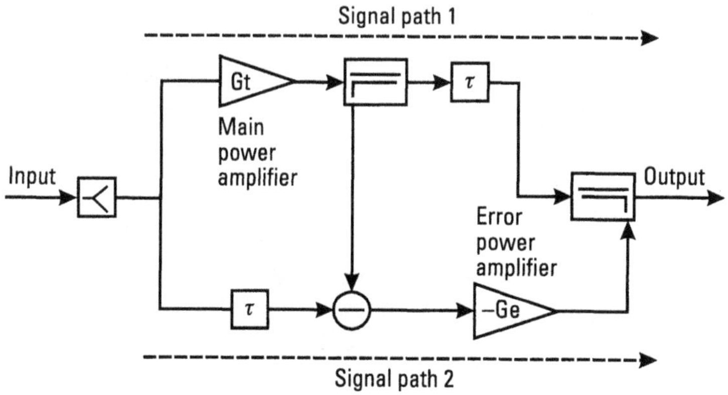

Figure 3.4 Feed-forward amplifier.

signal is split between the main signal path one and error signal path two. The high-power amplifier with gain *Gt* in the main signal path drives through a directional coupler; this device maintains good directivity and splits off a small amount of post amplifier power to the subtractor in the error signal path. The output of the subtractor is the difference between a delayed version of the input (and more linear) signal and the less linear amplified signal from the main signal path. With perfect components this difference signal only contains the distortion components added by the main power amplifier. The error amplifier inverts the difference signal and adds it to a delayed version of the main signal via the directional coupler at the output. Considering realistic system components, the final process will subtract more of the distortion components than the original signal, resulting in an output signal that more closely matches the input signal; this is equivalent to making the system more linear.

The feed-forward approach has the capability to reduce harmonic components by approximately 20 dB and provide an efficiency of approximately 10%.

The feed-forward design and other similar approaches can be implemented purely with analog components (transmission lines, amplifiers, and directional couplers) or by a mix of digital and analog components. More complex but better performing linearizers are possible using digital signal processing in the error signal path.

Table 3.3
Typical Wideband 2G and 3G Power Amplifiers

Manufacturer	Air Interface	Bandwidth	Power	Efficiency
WSI	CDMA 1,900 MHz	30 MHz	60W	8%
WSI	GSM 1,800 MHz	30 MHz	4 × 20W	10%
WSI	GSM 1,900 MHz	30 MHz	4 × 20W	10%
WSI	TDMA 1,900 MHz	30 MHz	60W	8%
WSI	UMTS 2,100 MHz	20 MHz	30W	12%
Wiseband	TDMA/CMDA/AMPS 800 MHz	25 MHz	40/80W	12%
Wiseband	GSM/CDMA/TDMA 1,900 MHz	60 MHz	40W	7%
Wiseband	UMTS 2,100 MHz	60 MHz	30W	8%

3.6.2 Power Consumption Efficiency

A significant trade-off for designers is the balance between wideband linear performance and power consumption. Table 3.3 lists basic parameters from two wide-band power amplifier manufacturers.

Multicarrier combining prior to the power amplifier eliminates the need for high-loss transmitter combiners prior to the antenna. The SDR design goal should be to use a wideband power amplifier that has better efficiency than the equivalent combination of single channel amplifiers and transmit combiners.

Figure 3.5 illustrates the form factor of a commercial wideband power amplifier suitable for 3G software defined radio.

Figure 3.5 Commercial wideband power amplifier. (*Source:* WSI, 2001. Reprinted with permission.)

3.7   Signal Processing Capacity Tradeoff

This chapter has concentrated on RF aspects that are driven by pure physics and the SDR requirement for multicarrier design. Many RF parameters are independent of the software running on the SDR platform, but not all.

A method for calculating the link budget has been provided; this implies that SNR can be maintained when base station and terminal separation is increased providing that the appropriate power is transmitted by both ends of the link. Given a homogenous transmission medium, increased power will improve the predetection SNR for 2G TDMA air interfaces. This is because the signal bandwidth is shared in time only. However, for CDMA systems the signal bandwidth is shared in power. The addition of a user to a CDMA cell will result in all other users increasing their power to equalize the apparent interference produced by the additional user. Therefore, for CDMA SNR is determined by both the link budget and the number of users in the cell.

Prior to a mobile terminal being switched on, neither the terminal nor the base station knows how far apart each is. The initial searching algorithms following switch-on need to search over a cell distance (or time) equal to the largest required distance. This searching process is performed by the baseband digital signal processing subsystem. The larger the distance the more computations that need to be performed and the greater the signal processing resources consumed.

Therefore, the SDR designer needs to tradeoff digital signal processing capacity with RF system parameters, such as cell distance/radius, power output, and the other link budget items. The *CDMA Systems Engineering Handbook* [16] is recommended for readers wishing to understand more about CDMA cellular network engineering. The book covers the IS-95 standard, but many of the examples and equations can be extended to the 3G CDMA technologies.

3.8   Design Flow

We have shown in this chapter how design decisions at one level or in one part of the system can impact other parts of the system. Figure 3.6 captures some of these decisions for a software radio development, with emphasis on the RF aspects. There will be iteration between each of the stages as detail is fleshed out and assumptions refined.

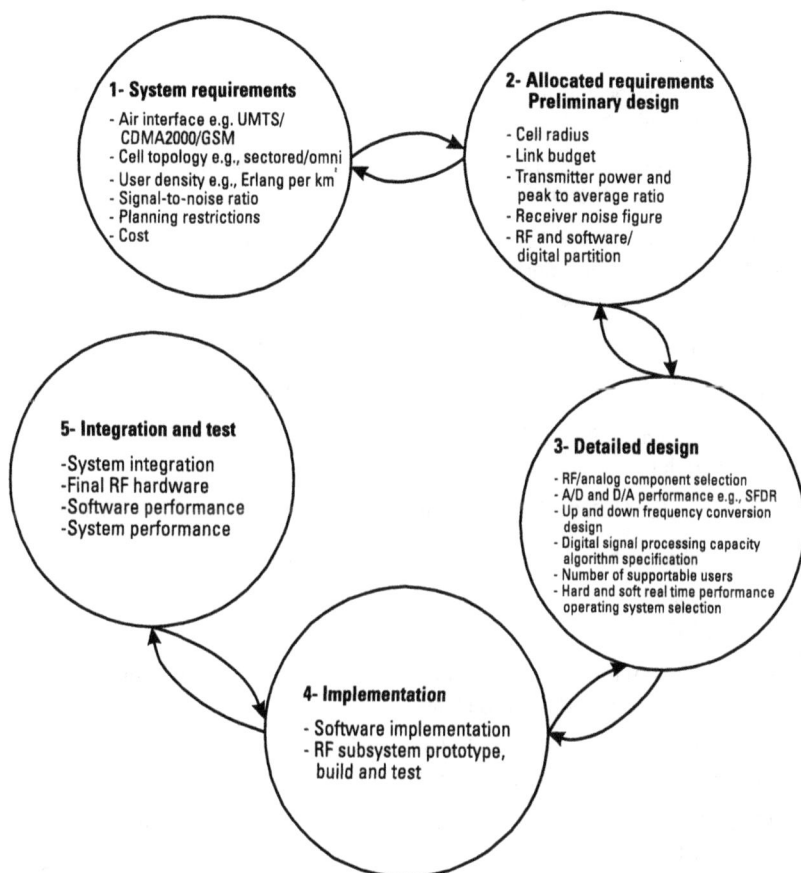

Figure 3.6 Design flow.

3.9 Conclusion

We have built upon the basic architecture of Chapter 2 and concentrated on RF issues for the 3G software radio designer. Significant RF development effort has been invested in key enabling technologies for software radio over recent years (e.g., wideband multicarrier power amplifiers).

The RF design is tightly coupled with the radio system design and the architectural partition between hardware and software. It is likely that the design of a generic platform for more than one air-interface will require careful system design and iteration to ensure the best outcome.

References

[1] ITU, "Recommendation ITU-R M.1034-1—Requirements for the Radio Interface(s) for Future Public Land Mobile Telecommunications Systems (FPLMTS)," 1997.

[2] ITU, "Recommendation ITU-R M.687-2—International Mobile Telecommunications-2000 (IMT-2000)," 1990, 1992, 1997.

[3] ITU, "Recommendation ITU-R M.1035—Framework for the Radio Interface(s) and Radio Sub-System Functionality for International Mobile Telecommunications 2000 (IMT-2000)."

[4] Lee, W., *Mobile Cellular Telecommunications*, New York: McGraw-Hill, 1995, Chapter 4.

[5] Yang, S., *CDMA RF System Engineering*, Norwood, MA: Artech House, 1998, p. 15.

[6] ITU, "Recommendation ITU-R M.1225—Guidelines for Evaluation of Radio Transmission Technologies for IMT-2000," 1997.

[7] TIA, "The CDMA 2000 ITU-RTT Candidate Submission (0.18)," July 27, 1998.

[8] ETSI, "Submission of Proposed Radio Transmission Technologies," January 29, 1998.

[9] Shanmugam, K. S., *Digital and Analog Communications Systems*, New York: John Wiley & Sons, 1979, p. 123.

[10] 3GPP2, "C.S0011 Recommended Minimum Performance Standards for CDMA 2000 Mobile Stations," http://www.3gpp2.org.

[11] 3GPP2, "C.S0010 Recommended Minimum Performance Standards for CDMA 2000 Base Stations," http://www.3gpp2.org, March 30, 2001.

[12] 3GPP, "3G TS 25.141 Third-Generation Partnership Project (3GPP) Base Station Conformance Testing (FDD)," http://www.3gpp.org, March 30, 2001.

[13] 3GPP, "3G TS 34.121 Third-Generation Partnership Project (3GPP) Terminal Conformance Specification—Radio Transmission and Reception (FDD)," http://www.3gpp.org, June 2001.

[14] Spear, J., and J. Crescenzi, "MCPA in 3G Apps," *Wireless Review*, June 1, 2000.

[15] Powerwave Technologies, "Multicarrier Power Amplifiers for W-CDMA Wireless Systems," 1998.

[16] Lee, J. S., and L. E. Miller, *CDMA Systems Engineering Handbook*, Norwood, MA: Artech House, 1998.

4

Analog-to-Digital and Digital-to-Analog Conversion

Software defined radios, as with all wireless digital communications devices, must at some point convert their discrete and digitized information stream into analog form for transmission. The organization of this book follows the software radio functions from the air-interface through to baseband. The previous chapter dealt with the analog functions from the antenna to the start of the digital domain. In this chapter we cover the important process of digital conversion, from analog to digital and digital to analog.

4.1 Introduction

In Chapter 1 we touched upon the progress that DSP devices have made since their inception in the 1980s. Possibly the weakest link in the SDR signal processing chain is the process of analog-to-digital conversion and, to a lesser extent, the process of digital-to-analog conversion. The Moore's Law principle is often referenced to highlight the rapid advances that DSPs and computing devices have made since their inception. Unfortunately, digital conversion techniques face more fundamental challenges; in many respects their progress has not kept pace with the computing devices that follow them in the signal processing chain. Digital conversion resolution improvements of approximately 1.5 bits every 6 years have been suggested [1]; however, the recent wireless boom and 3G focus have seen SDR-capable converters improve by 2 bits in approximately 2 years.

The first developments of wideband software defined 2G base stations in the mid-1990s were fundamentally limited by the resolution and spurious free dynamic range (SFDR) of the digital converters. Designs with bandwidths of 5 MHz or greater were generally limited to 12-bit ADCs; the SFDRs of these devices were considered insufficient for some air interfaces (e.g., GSM). We are now seeing suitable wideband 14-bit ADCs with enough performance for GSM, and better 16-bit converters are on the horizon.

4.2 Digital Conversion Fundamentals

Since digital conversion is such an integral function of the software radio, it is worthwhile to first review the associated fundamental digital signal processing concepts.

4.2.1 Sample Rate

Assume that we have a continuous time domain signal, y [(4.1)], where the frequency response of the signal is centered on 0 Hz and extends from $-B/2$ to $+B/2$ Hz.

$$y = f(t) \qquad\qquad (4.1)$$

Before any digital processing of the signal can take place, it must be sampled at discrete time intervals. The Nyquist sampling theorem states that the sampling frequency, Fs, must be greater than or equal to twice y's highest frequency component (i.e., $Fs \geq B$). Conversion of an analog signal into the digital domain by Nyquist sampling is also known as lowpass sampling. This is because all the frequency components from 0 Hz to Fs are recovered.

4.2.2 Bandpass Sampling

In Figure 4.1 we have a band-limited signal with bandwidth B that is centered on Fc Hz rather than 0 Hz as per the Nyquist example in Section 4.2.1. It is possible to choose a sampling frequency that is less than twice the highest frequency component ($Fc + B/2$) providing Fs is still at least $2B$. This is bandpass or undersampling; the technique takes advantage of the frequency domain repetition caused by the sampling process. To be practical the analog to digital converter must also have enough front-end analog gain and the

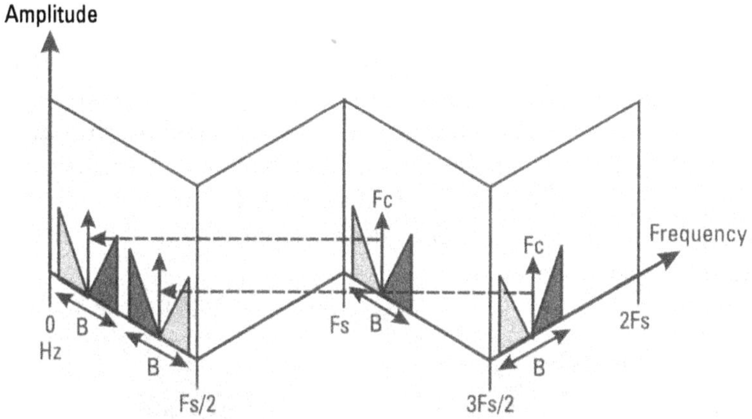

Amplitude

Figure 4.1 Sampling above the Nyquist frequency.

required linear performance extending beyond the top sampling frequency up to *Fc* + *B*/2. Figure 4.1 demonstrates how the Nyquist zones repeat themselves and how the spectrum beyond the Nyquist frequency folds back into the first Nyquist zone (*–Fs*/2 to *Fs*/2).

Figure 4.1 shows two relative positions for the band-limited signal centered on *Fc* Hz. The first position occurs when a relatively high sampling frequency is chosen and the signal falls between *Fs* and 3*Fs*/2. The second position is for a relatively lower sampling frequency, where the signal falls in-between 3*Fs*/2 and 2*Fs*. This figure shows that choosing a higher sampling frequency causes the signal to neatly fold back into the first Nyquist zone with the same frequency order as the signal had prior to sampling. Choosing the lower sampling frequency folds the signal back to the first Nyquist zone in reverse frequency order, with the low frequencies at the high end and vice versa. Frequency reversal can be corrected in the digital signal processing that follows sampling by inverting the quadrature component of the in phase (I) and quadrature (Q) signals.

Bandpass sampling has the advantage that the signal is also frequency downconverted in the same action, thereby reducing the need for an additional analog downconversion stage prior to sampling or a digital conversion stage following sampling. However, this approach assumes that the analog gain of the converter is not significantly diminished across the sampled signal's passband and that gain and phase distortion are within specification. Compared with oversampling where a lowpass antialiasing filter can be applied, bandpass sampling must use a bandpass filter.

4.2.3 Oversampling

A signal is considered oversampled when the sampling frequency exceeds the Nyquist frequency; in many cases this will be some integral multiple of the Nyquist frequency. When compared with bandpass sampling, the technique has the advantage of a simpler antialiasing filter (i.e., lowpass). However, oversampling has the disadvantage of a lower digital IF frequency, which may imply the use of nonstandard components and possibly an extra analog frequency conversion stage.

4.2.4 Antialias Filtering

Figure 4.1 demonstrates that all signals above $Fs/2$ are aliased or folded back into the first Nyquist zone, including all unwanted components and any noise. The analog to digital converter must be preceded by an analog antialiasing filter; this has the function of passing only the band of interest and rejecting all other unwanted frequency components.

The antialiasing filter must exhibit lowpass-band ripple, good phase linearity, and group delay performance with a minimum of insertion loss. It must achieve all these parameters with the steepest amplitude roll-off between the passband and the stopband.

For the basic architecture presented in Chapter 2, the antialiasing function is effectively performed by the entire analog frontend prior to the analog to digital converter. It is the combined amplitude and phase response of all the components that must be considered. Using two or more stages of IF analog conversion prior to the ADC can simplify the filtering design task.

In the future, as the analog section in the SDR reduces in size, the antialiasing job will become more difficult. For example, one could use a surface acoustic wave (SAW) filter as the prime antialiasing filter because of the design's excellent roll-off in the transition band. However, these SAW filters exhibit quite high insertion loss (e.g., 21.4 dB for the LG FS0140B1). This LG filter operates at the popular 140 MHz IF, has a 9.8-MHz bandwidth, a stop-band attenuation of 60 dB, and a transition band of approximately 1 MHz. One would need to ensure that the SAW filter was preceded by a very low noise amplifier to ensure adequate noise figure.

4.2.5 Effective Number of Bits

Although an ADC may convert the analog sample into an n-bit digital word, the process of making that conversion is prone to error and not all of the bits will be meaningful. The effective number of bits (ENOB) specifies the dynamic performance of an ADC at a specific frequency, amplitude, and sampling rate relative to an ideal ADC's quantization noise. ENOB is calculated by ENOB = (SINAD −1.76 dB)/6.02 dB [2].

4.2.6 Quantization

Sampled analog signals are quantized in time; to allow manipulation of sampled signals by computers and DSPs they must also be quantized in amplitude. The time-related quantization step size is simply the reciprocal of the sampling frequency Fs. The amplitude-related quantization step size q of an analog to digital converter is calculated by:

$$q = \frac{V_{fs}}{2^N - 1} \approx \frac{V_{fs}}{2^N} \tag{4.2}$$

where V_{fs} is the full-scale voltage range of the ADC and N is the number of binary bits. For example, a typical 14-bit ADC [3] suitable for SDR has a peak to peak full-scale voltage range of 2.2V. The AD6644 [3] will quantize the 2.2-V range into 16,383 equally spaced values resulting in a q of 0.134 mV. Assuming that the ADC rounds to the nearest quantization value, the worst-case error is $\pm q/2$.

Let's now consider an analog input sinewave with a full-scale peak to peak amplitude of A; the quantization step size is now:

$$q = \frac{A}{2^N} \tag{4.3}$$

For this case the quantization error, e, is assumed to be a uniformly distributed random variable in the interval of $\pm q/2$ with zero mean. The variance, or quantization noise power, is given by:

$$\sigma_e^2 = \int_{-q/2}^{q/2} e^2 P(e)\, de = \frac{1}{q} \int_{-q/2}^{q/2} e^2\, de = \frac{q^2}{12} \tag{4.4}$$

The signal to noise ratio can be calculated by dividing the average signal power in the sinewave, $A^2/8$, by the quantization noise power and expressing this in decibels, as follows:

$$SNR = 10\log\left(\frac{A^2/8}{q^2/12}\right) = 10\log\left(\frac{3 \times 2^{2N}}{2}\right)$$

$$= 6.02N + 1.76 \text{ dB}$$

(4.5)

For the 14-bit ADC example the theoretical SNR for an input sinewave will be 86 dB. Practical limitations in the production of high-performance converters make this theoretical figure harder to obtain as the number of bits is increased. As an example, a 10-bit ADC (AD9070) has a worst-case typical SNR of 55 dB compared with the (4.5) derived figure of 62 dB, while a real 14-bit ADC (AD6640) provides 73.5 dB of SNR compared with the theoretical figure. Digital to analog converters suffer the same problem (e.g., the AD9772 14-bit DAC achieves an SNR of 71 dB compared with the 86-dB theoretical figure).

For wide bandwidth converters and SDR applications the full Nyquist bandwidth SNR figure can sound deceptively poor. A more useful figure for SNR considers only the noise power in the information bandwidth of concern. For 3G modulation schemes the largest information bandwidth F_I, is 5 MHz (correlates to 3.84 Mcps UMTS). So for a bandpass information signal (at some IF) and a converter with a sampling frequency and Nyquist bandwidth, of F_s, the information band SNR_I is:

$$SNR_I = SNR_N + 10\log\left(\frac{F_s}{2F_I}\right)$$

(4.6)

where SNR_N is the full Nyquist SNR. SNR_I for a theoretical 14-bit converter such as the AD6640 using an Fs of 65 MHz will be 94.2 dB over a UMTS carrier bandwidth of 5 MHz.

4.2.7 Static and Dynamic Errors

Distortion produced by a digital converter can be attributed to its static nonlinearity and dynamic nonlinearity features. The converter's dc transfer function is used to characterize static nonlinearity and express this as a number of bits of integral nonlinearity (INL) and differential nonlinearity (DNL) error (e.g., AD6644 guarantees an INL of ±0.5 bits and a DNL of ±0.25 bits). The best straight line method of determining the INL error plots a line of best fit through the converter's dc transfer function and measures the difference between this and the actual transfer function. DNL is measured by taking the difference between the actual step width and the ideal value of one least significant bit (1 LSB).

The static parameters of INL and DNL are not the most important when comparing converters for use in SDR; they are more meaningful when considering applications in high-resolution imaging (e.g., a CCD digitizer).

Dynamic nonlinearity is the more important SDR-related parameter, and this is commonly expressed through specifications such as signal to noise and distortion (SINAD) and SFDR.

4.2.7.1 SINAD

The SINAD ratio is computed by taking the ratio of the rms-required signal power to the rms of all the unwanted spectral components (including harmonics) but excluding the dc component. A good data converter will exhibit a SINAD very close to its SNR (e.g., the AD6644 data sheet quotes the same SNR and SINAD figures up to 30.5 MHz where SINAD is 73 dB and SNR is 73.5 dB).

4.2.7.2 SFDR

SFDR is the ratio of the rms signal amplitude to the rms value of the peak spurious spectral component. The spurious component may or may not be a related harmonic of the input signal caused by nonlinearity in the converter.

SFDR is an excellent metric because its encapsulates linearity and quantization noise performance. It is a more important parameter than SNR or SINAD when considering the performance of either an ADC or DAC for SDR.

Spurious components have very narrow bandwidths and usually their amplitude is much larger than the rms quantization noise power within the information bandwidth. For a receiver the dynamic range will be limited by the spurious energy as the small information signals compete. In a wideband receiver the coherent nature of the spurious components can cause mixing with strong out-of-band signals producing unwanted in-band intermodulation products.

The quantization noise generated by a DAC within the information band is generally not important for the transmitter in a radio system. This noise is attenuated during propagation and ends up being significantly less than the thermal noise seen at the receiver. However the out of information band spurious components will be radiated and can cause unwanted intermodulation products appearing inside the information band of the receiver at the other end of the link.

4.3 ADC Techniques

In this section we discuss successive approximation and multipass conversion techniques and look at techniques to improve performance. Although successive approximation is significantly more common than multipass, it is generally not a feasible software radio ADC.

4.3.1 Successive Approximation

The successive approximation method of converting analog signals to N-bit digital signals is illustrated in Figure 4.2, which shows an example 8-bit ADC.

The main blocks for this method include sample and hold, successive approximation register (SAR), digital to analog converter, and a comparator circuit. Prior to each conversion the bits of the SAR are set to zero; the output of the DAC is, therefore, zero volts. Progressively and starting with the most significant bit, d_7, each bit of the SAR is set to one. The output of the

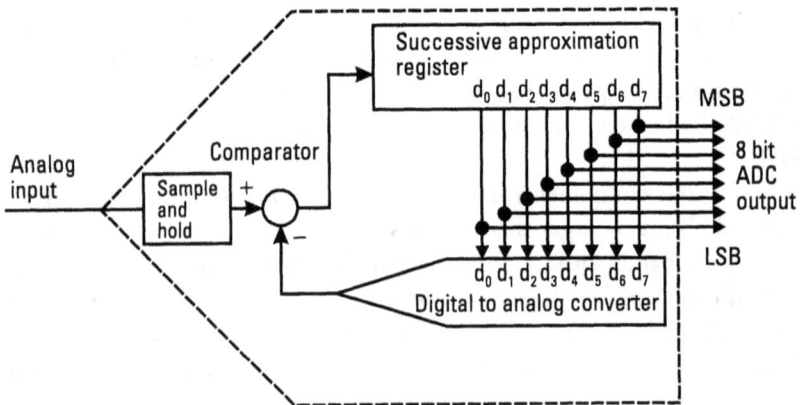

Figure 4.2   Eight-bit successive approximation ADC.

DAC is compared with the output of the sample and hold; the bit is left set if the input voltage exceeds the DAC voltage—otherwise, it is returned to zero. This process is continued until all the bits of the SAR have been tried; after eight clock cycles the digital word appears on the ADC output pins.

This technique uses the divide and conquer approach by searching half of the quantization levels for each iteration until the solution is found. The approach has the advantage of requiring a minimum of precision hardware; this is traded off for conversion speed, which is slow when compared with other techniques.

Successive approximation has been popular for more than 20 years (e.g., ADC0808 [4]: a 100-kHz, 8-bit device).

4.3.2 14-Bit Software Radio ADC

Successive approximation is not a feasible solution for software radio ADCs. A 14-bit, 65-MSPS implementation would need to be clocked at more than 14 times the sample rate, or 910 MHz, clearly an impossible task with current day technology.

ADCs with sufficient performance are now available for software receivers using the AMPS, IS136, IS–95, GSM, CDMA2000, or UMTS standards. The AD6644 [3] is a 14-bit device using the multipass architecture; see Figure 4.3.

The multipass architecture has an element of the successive approximation technique; however, it is significantly more complex. Its main functional blocks include three ADCs, two DACs, five track and holds, two comparators, and an error correction circuit. Instead of using N successive iterations, the multipass architecture performs the conversion in three steps.

The first stage, ADC1, converts the input analog sample into a 5-bit digital word and feeds this to the very precise 5-bit DAC1. Fourteen-bit precision is required for DAC1; this is achieved by expensive laser trimming. The first difference signal (Ain-ADC1) is produced by subtracting the 5-bit result from the input sample. The process is repeated to produce a second difference signal (Ain-ADC1-ADC2); the second DAC2 only requires 10-bit precision and does not need to be laser trimmed. At each stage appropriate delays are achieved by the various track and hold devices. Finally, the second difference signal is converted to 6 bits by ADC3.

The three ADCs produce 16 bits; these are fed to an error correction circuit, which produces a 14-bit result guaranteed to have no missing codes. If the input of an ADC is swept uniformly through the entire input range, it

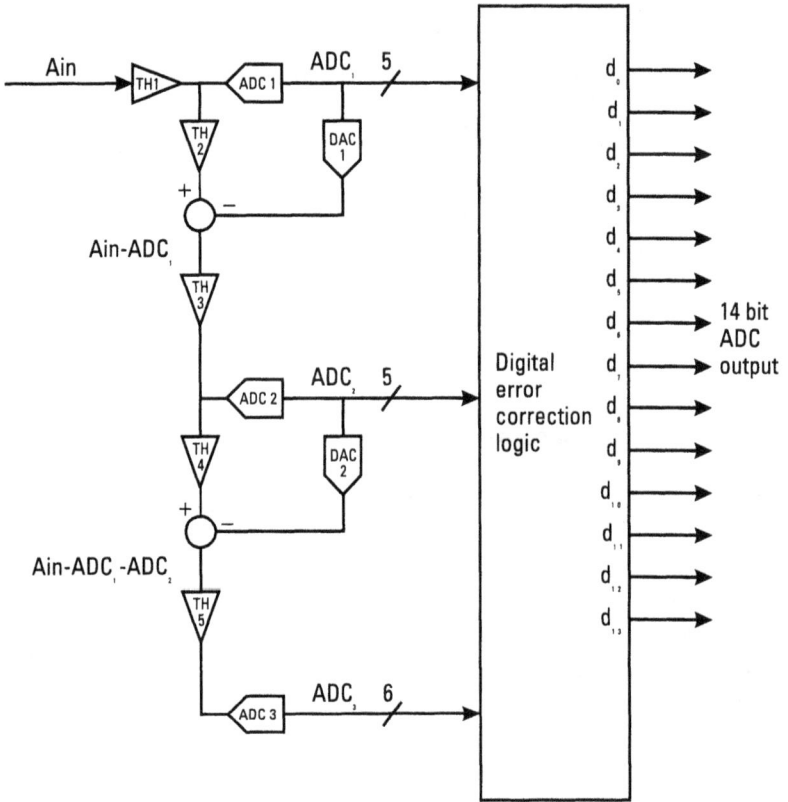

Figure 4.3 Multipass ADC.

is possible that one or more output codes may not occur—that is, one or more codes may be skipped or missing. Missing codes are most often caused by manufacturing errors or drift problems within the converter.

4.3.3 Dithering

Dithering is performed by adding an uncorrelated analog signal to the desired signal at the input to the analog to digital converter. The dithering signal is most often pseudorandom noise. Figure 4.4 illustrates a typical spectral plot of a 15.5-MHz sinewave sampled at 64 MHz by a high-precision 14-bit ADC without dithering. The Nyquist band consists of many spurious components between −90 and −100 dBFS; these are the result of quantizing the input sinewave.

By adding wideband pseudorandom noise to the input sinewave prior to sampling and digitization, the level of the spurious components can be reduced by approximately 20 dB. Figure 4.5 shows the added wideband noise near zero hertz with a mean level of –95 dB. Dithering randomizes the dynamic nonlinearity errors, effectively reducing the harmonic content of the error signal. In other words, the energy in the spurious signals is smeared out over many frequencies.

Figure 4.5 also demonstrates that even though the wideband dithering signal has increased the SFDR, noise energy has still been added to the system. The noise floor has increased as a result of the dither energy and the harmonic content of the error signal spreading out over many frequencies. This overall increase in noise floor (and decrease in SNR) may be acceptable if SFDR is the prime system requirement. Figure 4.6 illustrates a simple dithering circuit.

Dithering can be performed with a reduced impact on the noise floor by employing some different techniques. The first approach uses larger amplitude but narrower band pseudorandom noise. This narrowband dithering energy can be added to a section of the Nyquist band seldom used by communications systems, either centered on 0 hertz (dc) or the Nyquist frequency. Filtering can then be used to remove the dithering energy and restore

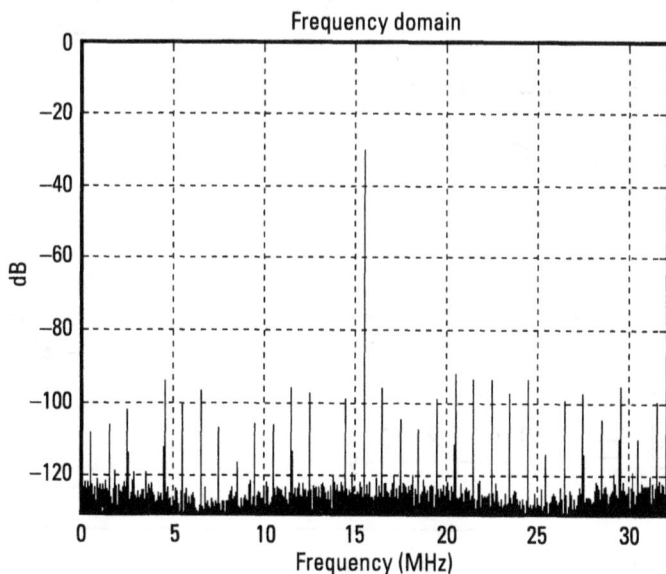

Figure 4.4 Sampled sinewave without dithering.

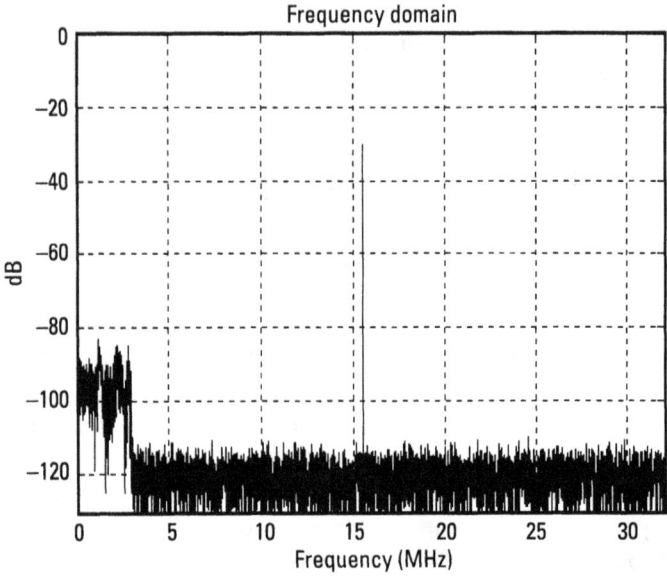

Figure 4.5   Sampled sinewave with dithering.

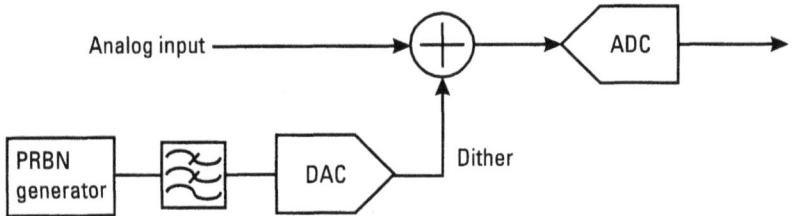

Figure 4.6   Dithering circuit.

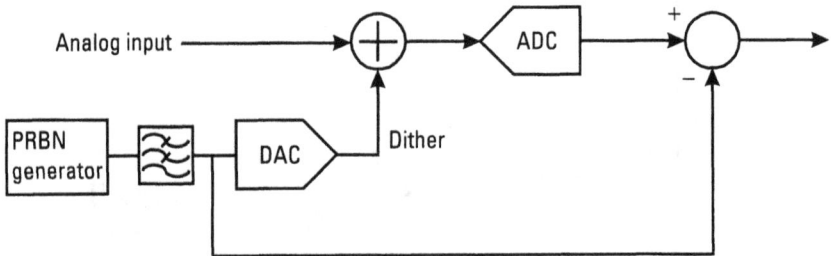

Figure 4.7   Digital subtraction of the dithering signal.

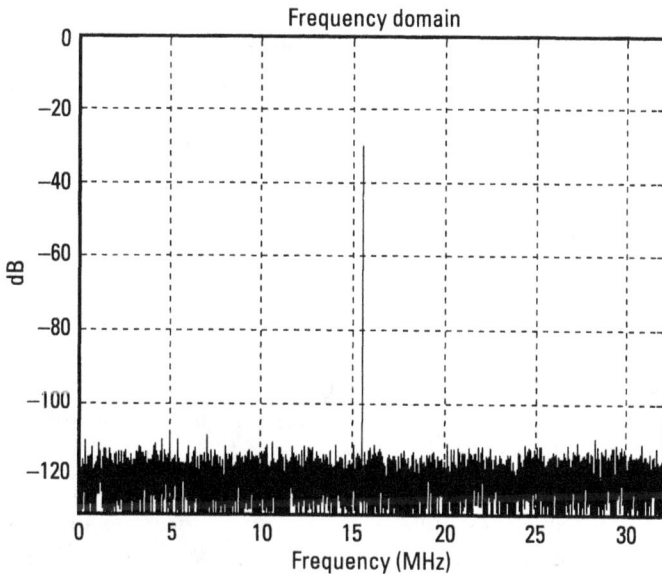

Figure 4.8 Digitally removed dither.

the SNR. A more complicated alternative digitally subtracts the pseudorandom dithering signal following digital conversion; see Figure 4.7.

Figure 4.8 illustrates the spectral content for the digitally removed dither.

4.3.4 Clock Jitter and Aperture Uncertainty

Clock jitter and aperture uncertainty are two terms used to express the same problem. Clock jitter has the effect of increasing system noise, increasing the uncertainty in sampled phase, and is a contributor to downstream problems such as intersymbol interference.

Basic analysis of ADCs and DACs assumes that the clocks used for sampling and holding are perfect. Typical clocks are square waves locked to highly stable sinusoidal reference sources; in some cases the reference is also locked to global time (e.g., GPS). A specified amplitude of either the rising or falling edge of the clock triggers the sampling and holding process. Natural phenomena, including bandwidth limitations, skew, and noise, prevent these clocks from being perfect. These effects cause the time interval between successive clock edges to vary or jitter. This variation in timing is random with a mean and variance.

For traditional nonspread narrowband modulation schemes (e.g., GMSK and GSM), relatively small amounts of clock jitter can result in considerable phase modulation and SNR reduction of the resultant ADC output. When a narrowband signal source is sampled at frequency F_s by an ADC with an rms sample clock timing jitter of σ_τ the phase modulation introduced in the source by the sampling process limits the SNR [5] to:

$$SNR = \frac{1}{\left(2\pi F_s \sigma_\tau\right)^2} \tag{4.7}$$

Equation (4.7) demonstrates why jitter can be a big issue with SDR and why it must be carefully considered. The aim of the software radio designer is to push the sampling frequency as high as possible to maximize the processable bandwidth of each processing chain. Higher intermediate and sampling frequencies cause a commensurate decrease in SNR (i.e., a doubling in sampling frequency will reduce the SNR by 6 dB for a fixed jitter). A value of σ_τ in the range of 1 picosecond has been suggested [6] as suitable for IF sampling. Considering a sampling frequency of 70 MHz and an rms jitter of 1 picosecond, SNR will be limited to 67 dB. Therefore, if the system used a 14-bit ADC with the values of F_s and σ_τ, SNR would be limited by the effects of jitter rather than the theoretical SNR performance due to quantization noise.

Because (4.7) only deals with the ratio of source signal power to error power introduced by the clock jitter, it cannot be directly applied to the spread spectrum case. A proper application of the theory must extend into the despreading process, since this is where the effects of clock jitter in direct sequence spread spectrum systems are manifested.

Stewart [5] provides an analysis of the fractional impacts of jitter on SNR for the IS–95 CDMA system. The fractional receiver SNR, Γ, is calculated by dividing the magnitude of SNR for the jittered case, SNR_J, by the magnitude of the SNR for the unjittered case, SNR_U:

$$\Gamma = \frac{\left|SNR_J\right|^2}{\left|SNR_U\right|^2} \tag{4.8}$$

and expressing this in terms of F_s and chip rate, f_c:

$$\Gamma = 1 - \left[\frac{F_s}{f_c} \right]^2 \sigma_\phi^2 \tag{4.9}$$

where σ_ϕ is clock jitter in radians.

For an IS-95 chip rate of 1.2288 Mcps and a sampling frequency of 70 MHz, if the SNR loss is to be limited to 0.1 dB, the clock jitter must be less than 2.6 milliradians or 6 picoseconds. Extending this analysis to UMTS and a chip rate of 3.84 Mcps, the clock jitter must be less than 1.9 picoseconds. Jitter and phase noise of this magnitude are possible with careful reference oscillator design.

4.3.5 Figure of Merit

When there are so many parameters that describe a functional element in a system, it can be useful to define a single parameter as a figure of merit (FOM). The FOM is then used to narrow down the scope of potential solutions for the software radio problem at hand.

For ADCs and DACs a useful FOM is sample rate multiplied by SFDR. As manufacturers do not always measure or quote SFDR for their converters, the surrogate FOM of sample rate multiplied by bits of resolu-

Table 4.1
Commercial High Speed ADCs

Manufacturer	Part Number	Bits	Sample Rate (MHz)
Analog Devices	AD9244	14	65
Texas Instruments	ADS809	12	80
Analog Devices	AD9071	10	100
Analog Devices	AD9218	10	105
Analog Devices	AD10200	12	105
Analog Devices	AD9432	12	105
Analog Devices	AD6645	14	105
Analog Devices	AD9433	12	125
Analog Devices	AD9410	10	210
Analog Devices	AD9430	12	200

tion can be used with care. The latter method of FOM is used in Table 4.1 to list ten of the highest performing ADCs currently available.

Selecting the best ADC will depend on the application, so while the AD9430 may exhibit an excellent combination of resolution and wideband performance, at 80 dB its SFDR is less than many other narrower band converters (e.g., 100 dB SFDR for the AD6645).

4.4 DACs

Ten of the highest-performing DACs available on the market are listed in Table 4.2. Most of these devices are suitable for 2G and 3G multicarrier SDR transmitters. They all have more than adequate sample rate to produce a single wideband signal that covers an entire mobile transmit band. As in Table 4.1, SFDR figures are either not quoted for all converters or are specified for very restricted sets of conditions. As an example, the AD9777, which tops the table with FOM, specifies an SFDR of 73 dB; the next highest device is the TI DAC5675 with an SFDR of 69 dB. The Intersil ISL5961 specifies an SFDR range of 52 to 79 dB but also suggests that for GSM and EDGE this can be increased to 94 dB for a signal frequency of 11 MHz and a data rate of 78 MSPS considering a 20-MHz window and a 30-kHz measurement bandwidth.

Table 4.2
Commercial High-Speed DACs

Manufacturer	Part Number	Bits	Sample Rate (MHz)
TI	DAC902	12	200
Intersil	ISL5861	12	210
TI	DAC904	14	200
Intersil	ISL5961	14	210
AD	AD9751	10	300
AD	AD9753	12	300
AD	AD9773	12	400
AD	AD9775	14	400
TI	DAC5675	14	400
AD	AD9777	16	400

4.5 Converter Noise and Dynamic Range Budgets

In Chapter 3 we discussed the budgeting for noise generation by each of the analog system components in the RF transmit and receive paths. We will now consider the budgeting of noise for the ADC and DAC.

4.5.1 DAC Noise Budget

The noise budget for a radio's digital to analog converter is largely driven by the required transmitter out-of-band signal mask. Out-of-band signals are those unwanted noise and spurious components that fall outside the modulation channel bandwidth but inside the wideband software radio DAC Nyquist bandwidth.

Noise and spurious components will be generated by the DAC and the analog elements that follow in the signal processing chain, with the power amplifier most likely to be the biggest unwanted signal contributor.

The spurious emissions mask will depend upon the air interface standard. For UMTS [7] the specified spurious emission levels depend upon the output power of the transmitter. For UMTS base stations with a power output of greater than 43 dBm (20W), spurious responses ($SFDR_{spec}$) greater than 3.5 MHz from the carrier must be less than 71 dB below the carrier power. For CDMA2000 base stations [8] with a power output greater than 33 dBm (2W), spurious responses greater than 1.98 MHz from the carrier must be less than 60 dB below the carrier power. The UMTS and CDMA-2000 spurious measurements assume a 30-kHz bandwidth.

A suitable design method starts with the noise and spurious components contributed by the DAC; the remaining margin is allocated to the transmitter's analog functional elements, with a few decibels set aside for an implementation margin. For wideband SDR applications the DAC's noise power, P_n, will be spread over the full Nyquist bandwidth. Let's assume that the sampling frequency is F_s and the fractional bandpass measurement bandwidth is F_m. The fractional noise power, P_m, in a fractional measurement bandwidth will be smaller and calculated by:

$$P_m = P_n - 10\log\left(\frac{F_s}{2F_m}\right)$$

(4.10)

Assuming that the full-scale required signal output is 0 dBm, the total noise power for a real 14-bit DAC over the Nyquist bandwidth will be approximately –73 dBm. Let us also make the DAC output of 0 dBm be the reference point (i.e., equivalent to 0 dBc), and we will ignore the gain through the rest of the transmit system. Then, if the sampling frequency is 70 MHz and the measurement bandwidth is 30 kHz, the fractional noise power will be –107 dBm.

Therefore, considering that typical high-performance software radio DACs exhibit SFDRs mostly in the range of 60 to 80 dB, it will be the spurious power, P_s, that falls inside a 30-kHz bandwidth that will dominate over the noise power due to the effects of quantization, jitter, and thermal noise.

Now consider that the system's DAC is only followed by analog components that perform the final frequency upconversion and power amplification. The mixers and amplifiers that make up this analog section will have a noise figure, F_{tx}. Therefore, the noise power produced by the DAC will increase to $P_m + F_{tx}$. Assuming a F_{tx} of 5–15 dB, the noise at the output of the power amplifier will be in the region of –102 to –89 dBc.

Assuming that a 30-kHz band is likely to contain at most one worst-case spurious component with power P_s, the total noise and spurious power must be less than the specified SFDR, as per (4.11).

$$10\log\left(10^{((P_m+F_{tx})/10)} + 10^{(P_s/10)}\right) < -SFDR_{spec} \qquad (4.11)$$

If $P_m + F_{tx}$ is 20 dB less than P_s, then (4.11) simplifies to:

$$P_s \leq -SFDR_{spec} \qquad (4.12)$$

Additional system imperfections that can contribute to the DAC's spurious outputs will be limited to effects caused by the nonlinear transfer function of the analog part of the system and mixing with the local oscillators. It is probable that the resultant second-, third-, sum-, and difference-order spurious components would not be any larger than their fundamental sources.

In the worst case DAC manufactures SFDR figures (e.g., AD9772A) and, considering (4.12), the 3G spurious emissions specification can be achieved for CDMA2000 (i.e., P_s is 12 dB above $SFDR_{spec}$, and for UMTS P_s is 2 dB above $SFDR_{spec}$).

4.5.2 ADC Noise Budget

Noise is important in the receive system, since its presence limits detectability. One of the toughest mobile detectability requirements in the presence of adjacent channel signals is the GSM 900 system [9]. Figure 4.9 graphically presents the GSM 900 receive signal environment [10] for a base station.

GSM requires a carrier to noise and interference ratio of 9 dB to meet the specified bit error rate. Therefore, the total spurious and noise power, P_{nt}, appearing in the GSM channel bandwidth of 200 kHz must be less than −110 dBm. The out-of-band +8-dBm signal will be attenuated by the wideband antialiasing filter, so the in-band −13 dBm adjacent channel sets the receiver's maximum signal level and reference.

Therefore, the thermal noise, N_{tx}, in the GSM band will be −121 dBm. If the noise figure of the analog front end is 5 dB, thermal noise at the input to the ADC will increase to approximately −116 dB. If the ADC full scale is set to −13 dBm, (4.6) tells us that the fractional ADC noise power, P_m, for a

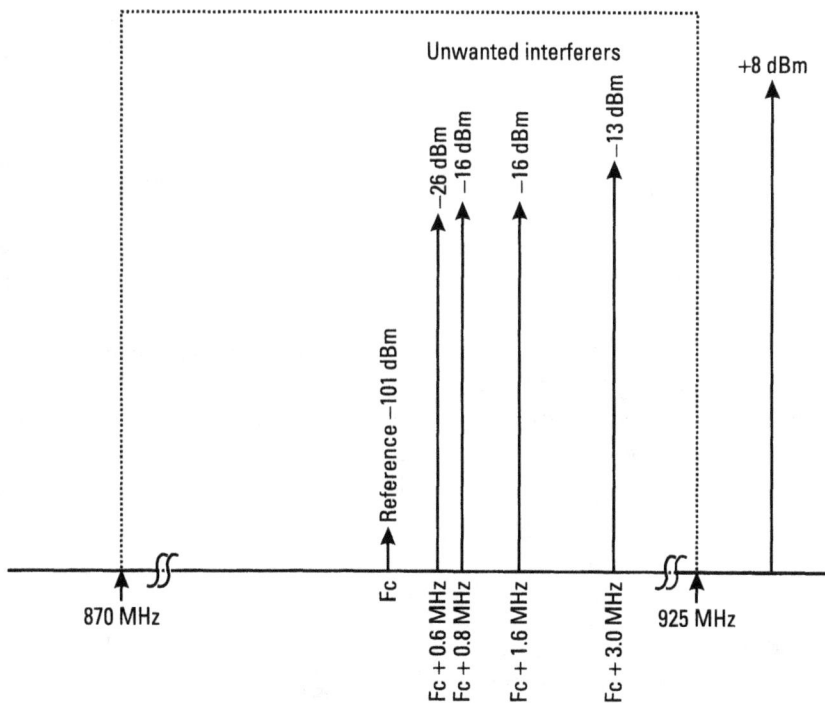

Figure 4.9 GSM blocking specification.

theoretical 14-bit ADC in the 200 kHz of a GSM channel will be approximately −121 dBm.

Assuming that a 200-kHz band is likely to contain at most one worst-case spurious component with power P_s, the total noise and spurious power must follow the equation:

$$P_{nt} < 10\log\left(10^{(P_m/10)} + 10^{(N_{tx}/10)} + 10^{(P_s/10)}\right) \tag{4.13}$$

Assuming an ADC SFDR of 100 dB, the spurious power will be −113 dBm. Also assuming no other sources of noise and spurious, P_{nt} will be approximately −110.8 dB; this gives a very small 0.8 dB implementation margin to meet the GSM900 specification. The situation could be improved by narrowing the IF bandwidth to 5 MHz and setting the maximum receiver level to −16 dBm (note that this solution is implemented in Table 8.5; see GSM 900 SDR Front End BW).

The assumption of one spurious component in the measurement band for the ADC and DAC is a statistical one; a more thorough analysis would consider the Monte Carlo simulation, and final validation of a design would be by prototyping and measurement.

4.6 Conclusion

Wideband high sample rate converters have recently matured to the point where they can be safely incorporated into the IF section of a 3G multicarrier software defined radio. Chapter 5 follows the processing chain back to baseband by covering digital up and downconverters. Although these devices source and sink digital signals, they also produce spurious signals and contribute to the imperfections added by each element in the processing chain. A complete systems analysis must, therefore, consider this effect and budget performance to each signal processing element.

References

[1] Winter, J., "Propagation Aspects for Smart Antennas in Wireless Systems," *IEEE Antennas and Propagation Symposium*, Volume 1, 2000.

[2] Walden, H., "Performance Trends for Analog to Digital Converters," *IEEE Communications Magazine*, February 1999.

[3] Analog Device, "AD6644 Data Sheet," http://www.analog.com.

[4] National Semiconductor, *1981 CMOS Data Book*, Victoria: The Dominion Press-Hedges & Bell, 1981, p. 3-3.

[5] Stewart, K., "Effect of Sample Clock Jitter on IF Sampling IS-95 Receivers," *IEEE*, 1997.

[6] Brannon, B., "Aperture Uncertainty and ADC System Performance," Analog Devices Application Note AN-501.

[7] 3GPP, "3G TS 25.141 Third-Generation Partnership Project (3GPP) Base Station Conformance Testing (FDD)," http://www.3gpp.org.

[8] 3GPP2, "C.S0010 Recommended Minimum Performance Standards for CDMA 2000 Base Stations," http://www.3gpp2.org.

[9] http://www.analog.com/publications/magazines/Dialogue/archives/29-2/wdbndradios.html.

[10] ETSI, "Radio Transmission and Reception GSM 05.05," Version 7.1.0, 1998.

5

Digital Frequency Up- and Downconverters

Digital upconverters (DUCs) and digital downconverters (DDCs) were introduced in Chapter 2; we now provide a complete coverage of the technology in recognition of its important place in the design of software defined radios for the cellular 2G and 3G marketplace. The main suppliers of DDC and DUC chips were Harris Semiconductor and GrayChip. Perhaps the importance of this technology is underscored by the fact that Harris sold its semiconductor business to Intersil in August 1999 and GrayChip was purchased by Texas Instruments in September 2001.

5.1 Introduction

DUCs and DDCs perform the digital frequency conversion function; from a software perspective they are a midway step between fully programmable DSPs and fixed-function ASICs. They are sometimes referred to as application-specific standard parts (ASSPs). Functionally they "sit" between digital baseband processing at the network end and digital conversion (ADC and DAC) at the RF end of the system. Most simply they can be considered as two-port digital devices, where wideband intermediate frequency signals exist at one port and narrowband single carrier baseband signals exist at the other. They are, however, not just simple devices that perform a fixed function once powered up; they have microprocessor interfaces and internal microcode processors.

5.2   Why Use DUCs and DDCs?

In Section 2.4.1 we looked at partitioning of the digital frequency conversion function and considered DSPs and ASSPs. Another potential choice for partitioning this function is the FPGA. Using these devices has the advantage of more closely fulfilling the software requirement (i.e., the SDR should be completely reconfigurable by software). However, an FPGA is likely to cost more in development time and production to implement. The gate count with an FPGA can be ten times that of a custom ASIC [1]; this will lead to the consumption of more power and board real estate for the same level of performance. Therefore, with DSPs discounted as too inefficient and FPGAs as too expensive, DDCs and DUCs are considered to have the best mix of software flexibility, performance, and cost for 2G and 3G mobile cellular radio.

5.3   Frequency Converter Fundamentals

The 2G and 3G multichannel digital frequency downconversion has the following major requirements:

- Filter or isolate a narrow band of frequencies (usually a modulated carrier) from the wideband source and reject the remainder of the band.
- Translate that isolated carrier down in frequency, usually from an IF to baseband.
- Reduce the data rate to some integer multiple of the information rate.

The 2G and 3G multichannel digital frequency upconversion has the following major requirements:

- Translate one or more narrowband signal sources (usually modulated carriers) up in frequency, usually from baseband to an IF.
- Combine the baseband sources to create one wideband result.
- Increase the data rate to a digital intermediate frequency rate.

We will now discuss the subfunctions needed to achieve these requirements while meeting the stringent 2G and 3G receive and transmit performance criteria.

5.3.1   Digital NCO

DDCs and DUCs contain on-chip digital NCOs. These oscillators generate very precise digital sine and cosine waveforms for use in the digital mixer, which performs the frequency translation function. Sine and cosine NCOs are used in quadrature I and Q mixing.

As the incoming (downconvert) or outgoing (upconvert) information signal is multiplied by the NCO-generated waveform, the NCO performance must be better than the specification for the multiplied result. Spurious responses in the NCO will mix with unwanted out-of-band signals; some of the resultant components will fall back in band and corrupt the required information signal. The SFDR of the NCO must be considerably better than the SFDR of the band-shifted and filtered output of the up- or downconverter. This is illustrated in Figure 5.1, which plots the spectral response for the 24-bit output of the NCO in an Intersil ISL5216 downconverter.

In Figure 5.1 the NCO was tuned to 40 kHz, Fs was 65 MHz, and the spectral estimation was measured by a 32-K point fast fourier transform (FFT), where the data was windowed using a Blackman function. The demonstrated SFDR was approximately 140 dB.

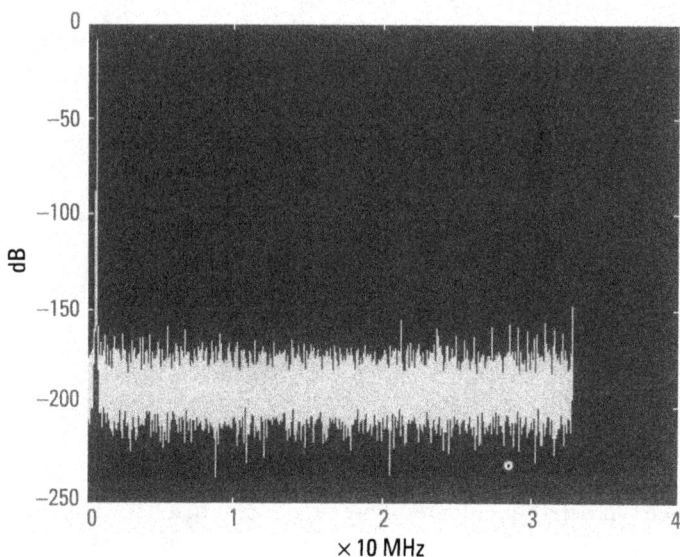

Figure 5.1   NCO spectral response. (*Source:* Intersil, 2001. Reprinted with permission.)

5.3.2   Digital Mixers

Digital mixers perform the same function as their analog counterparts. Sum and difference frequencies are generated by multiplying the incoming digitized source signal with a digital NCO. In most cases, for an upconverter, the input to the mixer will be quadrature (as provided by the baseband processing), and the output will also be quadrature, since the mixer is followed by several stages of quadrature filtering. For a downconverter the input will not be quadrature, since it is usually fed by a nonquadrature ADC; however, the output will be quadrature, since the mixer is again followed by several stages of quadrature filtering. The required output signal will either be the difference signal for downconversion or the sum signal for upconversion; the other component is not needed and will be removed by the following stages of digital filtering.

5.3.3   Digital Filters

Most digital filter designs are either infinite impulse response (IIR) type or finite impulse response (FIR) type. As the naming suggests, IIR filters have impulse responses that are infinite and FIR filters have ones that are not. A choice between the two can be made by matching the design requirements with the characteristics of the IIR or FIR type.

Digital FIR filters are linear systems and because they process sampled data they are also discrete in time. Linear systems follow the superposition principle, assume that the inputs to the system are $x_1(n)$ and $x_2(n)$ and the matching system responses are $y_1(n)$ and $y_2(n)$. Given an input of:

$$x(n) = a_1 x_1(n) + a_2 x_2(n) \tag{5.1}$$

where a_1 and a_2 are constants, the output of the system $y(n)$ is the straightforward summation of the individual responses:

$$y(n) = a_1 y_1(n) + a_2 y_2(n) \tag{5.2}$$

The superposition principle is very useful for the analysis of linear systems. If we can determine the response of a system to a basic waveform (e.g., step or impulse function) and then wish to discover the response to a more complex waveform (e.g., one that carries information), if the complex waveform can be expressed in terms of known simple waveforms, the response can be calculated by superposition of the responses to those simple waveforms.

5.3.3.1 IIR Filters

The major advantage of the IIR filter is that it generally can achieve the same magnitude response of an FIR filter by using a lower-order design (e.g., fifth-order FIR could equal a third-order IIR). A reduced order will mean less computations (multiplications and additions) and lower signal processing load.

The tradeoff for lower processing load it that it is impossible with an IIR to achieve a linear phase response, and filter implementation is not unconditionally stable. If distortion in the system's phase response can be tolerated and careful design is employed, an IIR can be a sensible choice, particularly for low-cost consumer devices, since the result may allow the use of a smaller, lower-cost DSP. A smaller (in processing capacity) DSP will usually consume less power and, for portable devices, increase battery life.

Since information is carried by both amplitude and phase in digital mobile communications systems, the IIR filter is generally not used; this is true for DDCs and DUCs.

The transfer function $H(z)$ of an $(N - 1)$th-order IIR filter can be expressed as the ratio of the numerator polynomial $B(z)$ and denominator polynomial $A(z)$ as follows:

$$H(z) = \frac{B(z)}{A(z)} = \frac{b_0 + b_1 z^{-1} + b_2 z^{-2} \ldots b_{N-1} z^{-(N-1)}}{1 + a_1 z^{-1} + a_2 z^{-2} \ldots a_{N-1} z^{-(N-1)}} \qquad (5.3)$$

The design requires $2N - 1$ filter coefficients with the input samples $x(n)$ and output samples $y(n)$ expressed as:

$$y(n) = -a_1 y(n-1) - a_2 y(n-2) \ldots - a_{N-1} y(n-(N-1))$$
$$+ b_0 x(n) + b_1 x(n-1) + b_2 x(n-2) \ldots + b_{N-1} x(n-(N-1)) \qquad (5.4)$$

This filter can be implemented using the direct form; a second-order example is illustrated in Figure 5.2. In practice the one sample delays (z^{-1}) would be implemented using RAM for a general-purpose signal processor or by a hardware shift register if the designer has chosen a DUC or DDC chip.

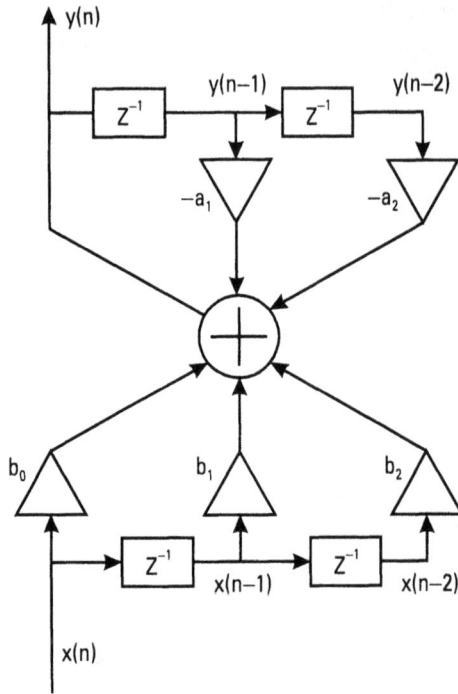

Figure 5.2   IIR filter realization.

The potential for instability is clearly demonstrated by the delayed feedback of the output samples $y(n)$ through the coefficients a_1 and a_2.

Apart from the direct form there are many other implementation choices, including cascade, parallel, frequency sampling, fast convolution, lattice, and wave.

5.3.3.2   FIR Filters

The major advantage of the FIR filter is that it is possible to achieve an exact linear phase response with unconditionally stability. For this reason it is the most popular choice of filter for use in digital communications. A recent search on "FIR Filter" resulted in 62,800 hits compared with 19,400 hits for "IIR Filter" using a popular Web search engine.

For an FIR filter the denominator polynomial $A(z)$ is equal to 1 and the transfer function $H(z)$ is equal to:

$$H(z) = \frac{B(z)}{A(z)} = \frac{b_0 + b_1 z^{-1} + b_2 z^{-2} \ldots b_{N-1} z^{-(N-1)}}{1} \qquad (5.5)$$

with the input samples $x(n)$ and output samples $y(n)$ expressed as:

$$y(n) = b_0 x(n) + b_1 x(n-1) + b_2 x(n-2) \ldots b_{N-1} x(n-(N-1)) \qquad (5.6)$$

Figure 5.3 illustrates the direct form of an example second-order FIR; unconditional stability is obvious by the lack of any feedback when compared with the IIR in Figure 5.2.

5.3.4 Halfband Filters

Halfband filters are a special case of FIR filter that find application where multirate processing is required. They are characterized by an odd number of coefficients (N), where every other coefficient is zero except $b_{(N-1)/2}$. The

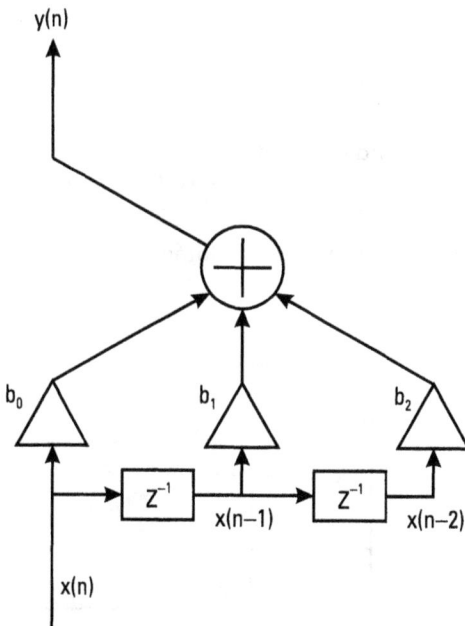

Figure 5.3 FIR filter realization.

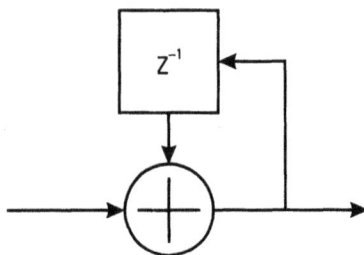

Figure 5.4 Basic integrator building block.

passband extends from zero to a quarter of the Nyquist band (i.e., half the available bandwidth is passed and the other half is rejected—hence, the name "halfband").

Halfband filters are popular in digital frequency converters because they require only half the coefficients of a standard FIR filter and because they ignore every second sample (they halve the output data rate, or decimate, by two).

5.3.5   Cascaded Integrator Comb (CIC) Filters

CIC filters are an important class of filter that efficiently performs decimation and interpolation [2, 3]. They are a flexible multiplier free filter suitable for implementation in hardware and can handle large rate arbitrary changes. Multipliers are complex and power-hungry functions to implement in silicon; for this reason the CIC is a popular choice.

The basic integrator and comb building blocks are illustrated in Figures 5.4 and 5.5, respectively.

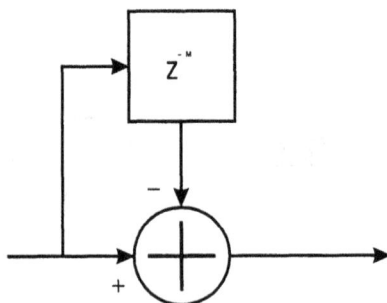

Figure 5.5 Basic comb building block.

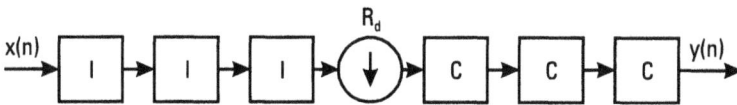

Figure 5.6 Three-stage decimating CIC filter.

An integrator is a single pole IIR filter with unity feedback; the function is also known as accumulation. Its amplitude transfer function is single order (-6 dB per octave) lowpass. However, the filter has infinite gain at 0 Hz and is therefore inherently unstable.

The comb is an odd-symmetric FIR filter where the Mth delayed sample is subtracted from the current sample; M can be any positive integer and is usually in the range of one to two. The other major parameter for the comb is the rate change, R. For $R = 1$ and $M = 1$ the amplitude transfer function is single order highpass.

CIC filters are constructed by cascading P integrators (I) or combs (C) followed by a decimator (see Section 5.3.6 for an explanation of decimation by R_d function) and then another P combs or integrators. Figure 5.6 illustrates a decimating CIC filter where $P = 3$.

Figure 5.7 shows that the decimating filter can be turned into an interpolating filter by swapping the integrators with combs and the decimator (rate reducer by R_d) with a zero stuffer (i.e., rate increaser by R_i; see Section 5.3.6 for zero stuffing explanation).

The combination of integrators and combs results in an unconditionally stable filter. For decimating implementations, integrator overflow is avoided by using two's complement arithmetic and by keeping the system gain to less than one. For interpolating filters, overflow is avoided by the comb stages and zero stuffing that precedes the integrators.

5.3.6 Decimation, Interpolation, and Multirate Processing

The requirement to process data at more than one sample rate in a system has long been recognized and is now referred to as multirate processing [4–8]. The two key multirate DSP operations of decimation and interpolation are simple yet efficient algorithms most often used to decrease or increase the data rate at any given point in a system.

The signal processing use of the word "decimation" does not accord with the dictionary definition of removing one in ten. It is instead used to indicate the periodic removal of all samples except every R_dth sample where

$$R_i$$

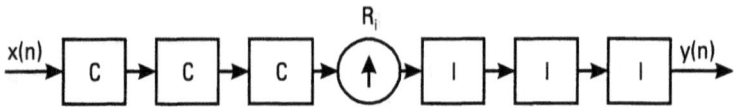

x(n) → [C] → [C] → [C] → (↑) → [↓] → [↓] → [↓] → y(n)

Figure 5.7   Three-stage interpolating CIC filter.

R_d is an integer. Decimation is generally only referred to if R_d is greater than one. The reduction in sample rate by a factor of R_d produces a reduction in bandwidth of the same factor. The circle with a down arrow in Figure 5.6 is a decimate by R_d function.

For a signal, $x(n)$, with a sampling frequency of F_s, interpolation increases the sample rate by a factor of R_i. This is achieved in three steps: zero stuffing, lowpass filtering, and gain. The zero stuffing function inserts $R_i - 1$ zero-valued samples between each input sample; this increases the data rate but produces aliased images of $x(n)$. The images are removed by lowpass filtering, and, because the energy in $x(n)$ is spread into these images and then lost by the filtering process, a gain stage is required to restore the original signal amplitude. The circle with an up arrow in Figure 5.7 is an interpolate by R_i function.

Wideband multicarrier software defined radio is an example of a multirate system. This is because the sampling frequency at IF is fixed for many information signals on different carrier frequencies. To move each carrier to baseband or 0 Hz, for example, each carrier must be shifted by a different increment relative to the sampling frequency. However, the baseband bandpass characteristic must be the same for each carrier, and the data rate must be an integer of the symbol rate. This can only be achieved by a mix of decimation and interpolation or multirate processing.

5.4   DUCs

The most recent 3G-capable digital upconverter from Intersil is detailed in this section.

5.4.1   ISL5217

The Intersil ISL5217 [9] is a quad programmable digital frequency upconverter. All four channels can be independently used for narrowband air interface applications such as IS 136, GSM, EDGE, IS-95, and CDMA2000-1x.

Two, three, or four channels can be combined together to support wider-band applications such as CDMA2000-3x-MC, CDMA2000-3x-DS, TD-SCDMA, and UMTS. Device limitations allow the ISL5217 to support three midband carriers (CDMA2000-3x-MC) or one wideband carrier (CDMA2000-3x-DS or UMTS). The device can output at a sample rate up to 104 MSPS with input rates up to 6.5 MSPS.

Figure 5.8 illustrates the ISL5217.

5.4.1.1   Data Input Routing

Data can be input through any one of four serial channels (Figure 5.8, SDA-SDD) or via the device μprocessor interface. Each of the upconverters four channels (Channel 0–Channel 3) can select any serial input as its source. The multiplexing arrangement at the front of the device also allows more than one channel to select its input from the same serial source. This would most likely be used in a smart antenna (see Chapter 9) application, where the output channels would contain the same information on the same carrier frequency, but the channel phase would be varied to steer a beam in the spatial domain.

5.4.1.2   Data Modulation

Baseband input signals can be vector modulated by using the shaping filter functional block followed by the gain profile block; these can be programmed for QASK for IS 136, EDGE, and IS-95. An FM modulation block is also provided; this can be set up either for band-limited mode, as required by AMPS/NMT, or for pulse shaping mode, as required by GSM. Two filters can be loaded simultaneously and switched dynamically via the μ Processor Interface for multistandard applications.

5.4.1.3   Sample Rate NCO

The sample NCO block is a 48-bit programmable oscillator that provides the clock and phase information to the data input FIFOs, shaping filters, and interpolation filters. The NCO has a coarse phase output, which drives the shaping filter, and a fine phase output, which drives the interpolation filter. The function includes a 32-bit leap counter; this is used to realize fixed integer interpolation rates and stop symbol slip by forcing the NCO phase back to its proper value. The quarter symbol required for GSM can be easily added by separately resetting the five MSBs to force the shaping filter phase to zero.

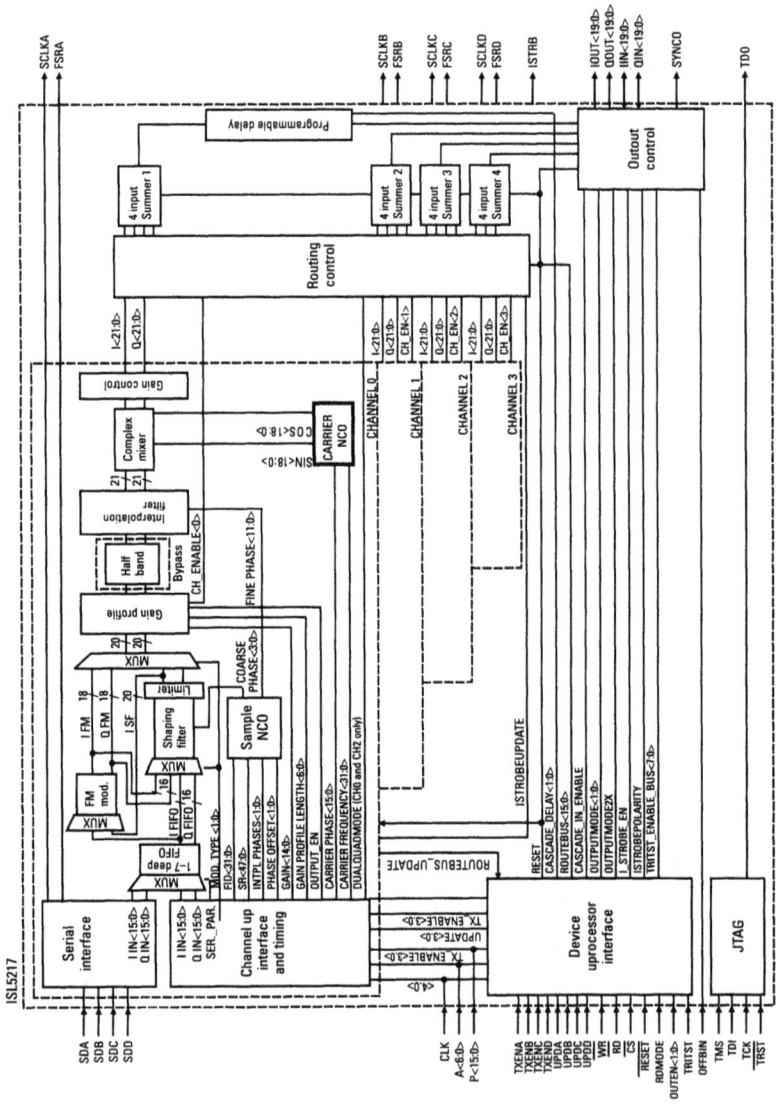

Figure 5.8 ISL5217 detailed block diagram. (*Source:* Intersil, 2001. Reprinted with permission.)

5.4.1.4 Shaping Filters

The shaping filter block contains an interpolating FIR filter; this can be used for pulse shaping. The I and Q channels each have their own 16 to 256 tap FIR filter. Memory is provided for each channel sufficient to store two sets of filter coefficients that can be dynamically switched during multimode operation; in this mode the stored FIR filters can have between 16 and 128 taps. The data format for the filter coefficients is a 24-bit number with a 20-bit mantissa and a 4-bit exponent. The I and Q channels of the FIR filter output are each 20 bits wide, and saturation logic takes care of peaks exceeding ±1.

5.4.1.5 Gain Profile Block

This function provides −0.0026 to −144 dBFS of programmable gain and transmit profile shaping control. The required profile is stored in a 128-bit by 12-bit RAM and is used to provide transmit ramp-up and quench fading to control the side lobe profile in burst mode. Gain profiling is controlled by the TXEN$\underline{X}$ signal at the μProcessor Interface and the function can be bypassed if required.

5.4.1.6 Halfband Filter

The ISL5217 Gain Profile function is followed by a halfband filter, designed to reduce interpolation images and improve the overall SFDR of the output signal. The filter is used in a fixed interpolate by two mode or it is bypassed. Interpolation is accomplished by zero stuffing and lowpass filtering using 11 fixed coefficients.

5.4.1.7 Interpolation Filter

This block provides the channel with its final filtering and resampling. Figure 5.9 illustrates the frequency response for the interpolating filter. It has the capability to output data at integer or noninteger interpolation rates to match the symbol rate to the final output clock rate.

5.4.1.8 Complex Mixer and Carrier NCO

All the ISL5217 functional blocks discussed so far process and produce baseband data. The complex mixer and carrier NCO provide the resources to perform the frequency upconversion where the result is a 20-bit I and Q output signal. The carrier frequency can be programmed over a 32-bit range with phase being programmable over a 16-bit range. Spectral inversion and

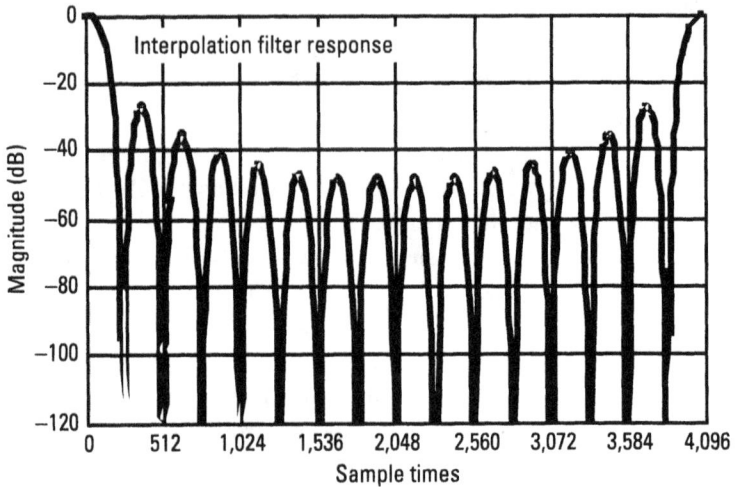

Figure 5.9   Interpolation filter response. (*Source:* Intersil, 2001. Reprinted with permission.)

vector rotation can be handled internally in the chip by programming the sign of the carrier frequency.

5.4.1.9   Gain Control

Unwanted changes in signal amplitude caused by the cascaded response of all the previous processing stages can be compensated for in the final gain control stage. This is implemented by using a scaling multiplier followed by a 3-bit scaling shift. Gain can be varied anywhere from –0.0026 to –144 dBFS.

5.4.1.10   Output Routing, Summing, and Control

This device provides flexible routing control, 4 input Summer, and output control blocks. The routing block allows each of the four channels the possibility of being connected to one input on each of the four summer blocks. The 4 input Summer has an additional input, which is used to put the ISL5217 into a cascaded mode. In this mode the fifth input can either accept data from an on-chip upconverter channel or from an off-chip channel.

An important part of wideband multichannel software radio is the combination at digital level of more than one carrier or channel for transmission by a single analog chain. The cascaded mode provides the designer with a solution to this requirement. Up to 16 narrowband channels (e.g., GSM) can be combined across four ISL5217s using this mode.

5.4.1.11 2G Performance

Examples of the performance for this upconverter using the GSM air interface are illustrated in Figures 5.10, 5.11, and 5.12 [9]. For these performance results the upconverter was set in cascade mode with four channels of continuous GSM transmission being combined and output. Figure 5.10 shows a 2-MHz window of the spectral output from an ISL5217 evaluation board; this board includes an HI5828 dual 12-bit DAC. A sample rate of 270.833 kHz was used together with a clock rate of 80 MHz.

The matching constellation diagram for this GSM example is plotted in Figure 5.11. The GSM phase accuracy specification of 5° rms was easily achieved (i.e., measured phase error was 0.2° rms).

The most important example of GSM performance is illustrated in Figure 5.12. The 20-bit cascaded output is presented as a full-band spectral envelope using a 32-K FFT. The results indicate an SFDR of 100 dB.

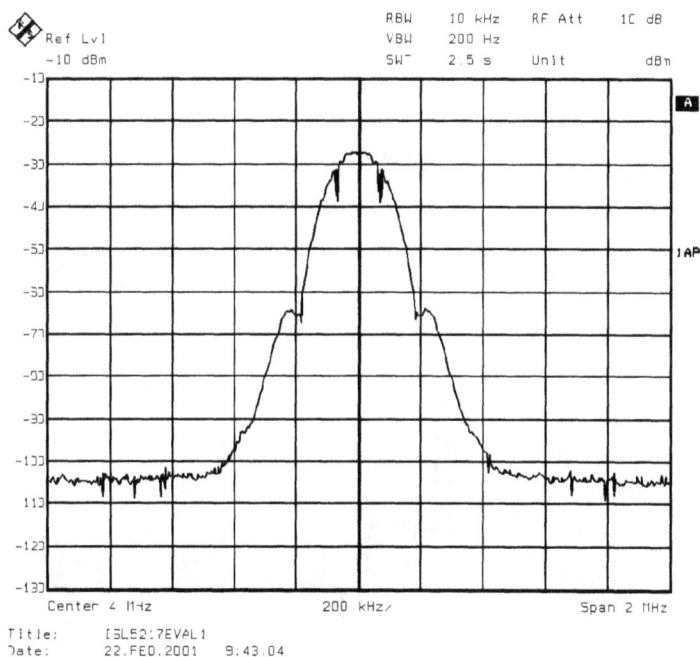

Figure 5.10 ISL5217 GSM analog spectrum. (*Source:* Intersil, 2001. Reprinted with permission.)

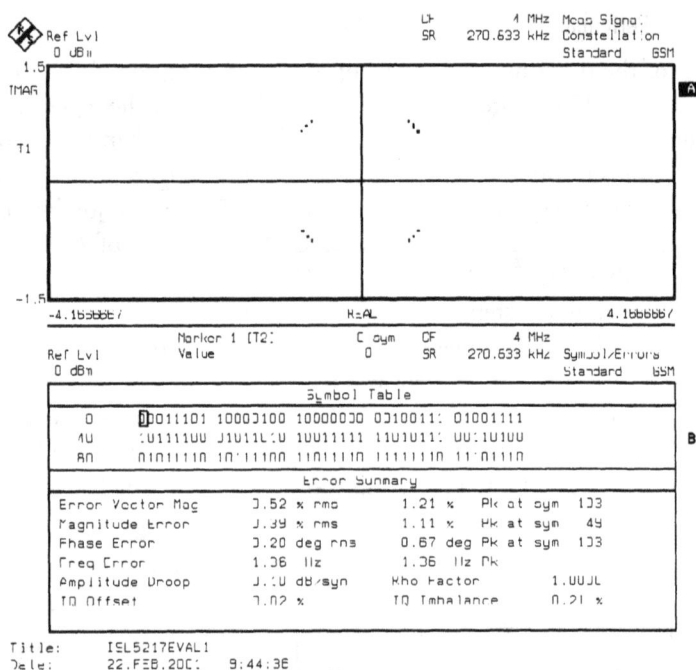

```
                                                    LF        4 MHz  Meas Signa:
      Ref Lvl                                       SR  270.633 kHz  Constellation
         0 dB                                                        Standard    GSM
    1.5
  IMAR                                                                               A

    T1



  -1.5
      -4.166666                          REAL                        4.166666
                   Marker 1 [T2:          C oym   CF         4 MHz
      Ref Lvl      Value                  0       SR   270.633 kHz  Symbol/Errors
      0 dBm                                                         Standard    GSM
                                  Symbol Table
         0      00011101 10003100 10000030 03100111 01001111
        AU      :U1111UU J1U11L:U 1UU11111 11U1U11: UU:1U1UU                          B
        BO      0101111n 1n 1110n 110111 1n 1111111n 11 0111n
                                  Error Summary
      Error Vector Mag        3.52 % rms      1.21 %   Pk at sym  133
      Magnitude Error         3.39 % rms      1.11 %   Pk at sym   49
      Phase Error             3.20 deg rms    0.67 deg Pk at sym  133
      Freq Error              1.36 Hz         1.36 Hz Pk
      Amplitude Droop         3.10 dB/syn     Rho Factor      1.0UJL
      IQ Offset               3.02 %          IQ Imbalance    0.21 %

  Title:     ISL5217EVAL1
  Date:      22.FEB.200:   9:44:38
```

Figure 5.11 ISL5217 GSM constellation diagram. (*Source:* Intersil, 2001. Reprinted with permission.)

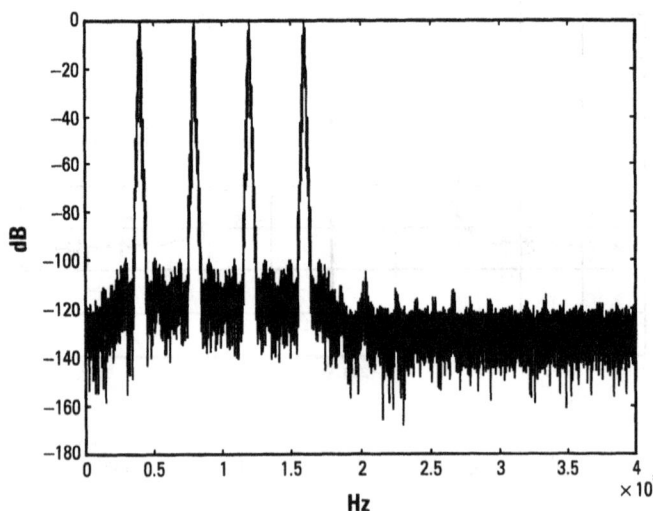

Figure 5.12 ISL5217 GSM multicarrier spectrum. (*Source:* Intersil, 2001. Reprinted with permission.)

5.4.1.12   3G Performance

In previous chapters we have identified UMTS and CDMA2000 as the major 3G technologies. CDMA2000 uses many air interface modes and is backward compatible with the 2G IS-95 CDMA standard. The three main modes are CDMA2000-1x (single carrier limited to the IS-95 channel bandwidth of 1.25 MHz), CDMA2000-Nx-MC (multicarrier mode with N adjacent 1.25-MHz carriers), and CDMA2000-Nx-DS (direct sequence mode uses larger spreading factor on a single carrier). The spectral output for CDMA2000-3x-MC is shown in Figures 5.13 and 5.14; after conversion to analog via a 12-bit DAC, the sample rate was 1.2288 MHz and the clock rate was 61.44 MHz.

Figure 5.14 indicates that the SFDR for CDMA2000-3x-MC was approximately 75 dB.

Similar conditions were used to produce the spectral plots shown in Figures 5.15 and 5.16 for the UMTS case. All four channels of the ISL5217 were required to produce this UMTS carrier at a center frequency of 7.3424 MHz; the clock rate was 61.44 MHz and the sample frequency was 3.84 MHz.

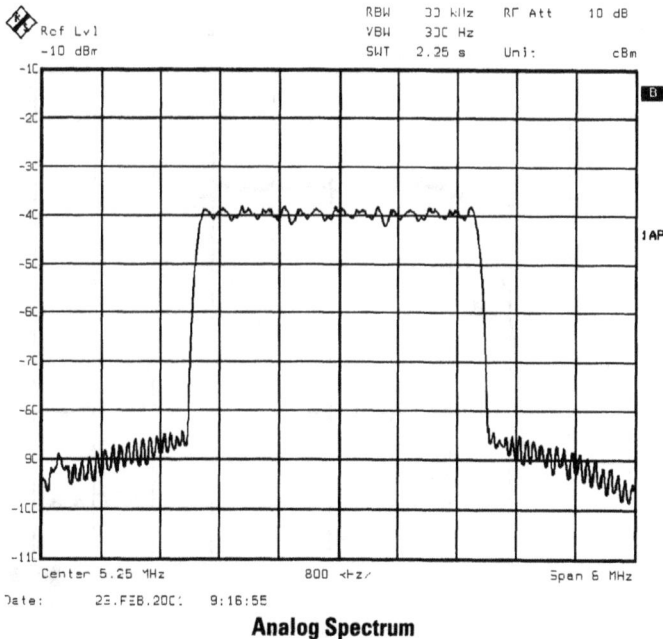

Analog Spectrum

Figure 5.13 ISL5217 CDMA2000-3x-MC analog spectrum. (*Source:* Intersil, 2001. Reprinted with permission.)

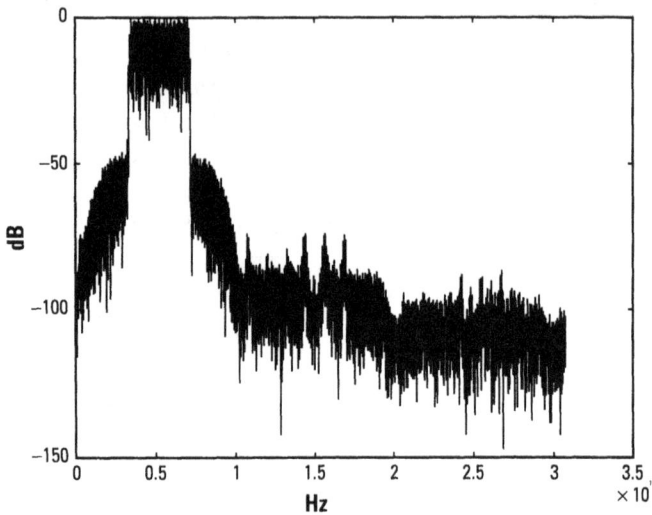

Figure 5.14 ISL5217 CDMA2000-3x-MC analog spectrum. (*Source:* Intersil, 2001. Reprinted with permission.)

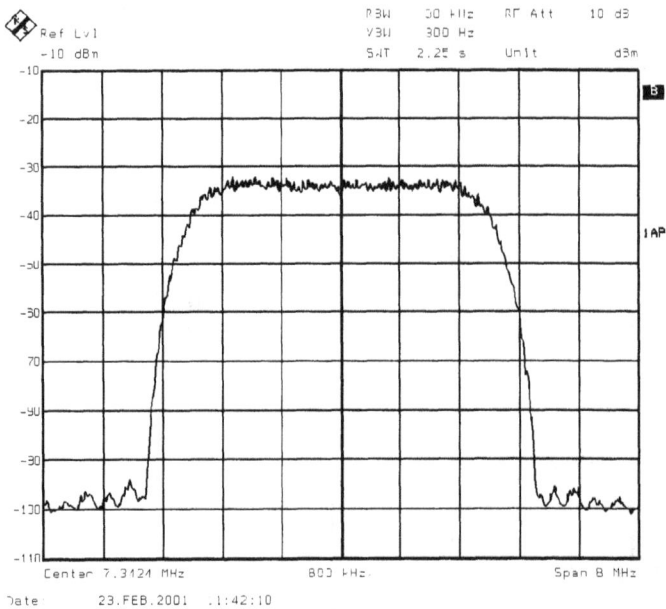

Figure 5.15 ISL5217 UMTS analog spectrum. (*Source:* Intersil, 2001. Reprinted with permission.)

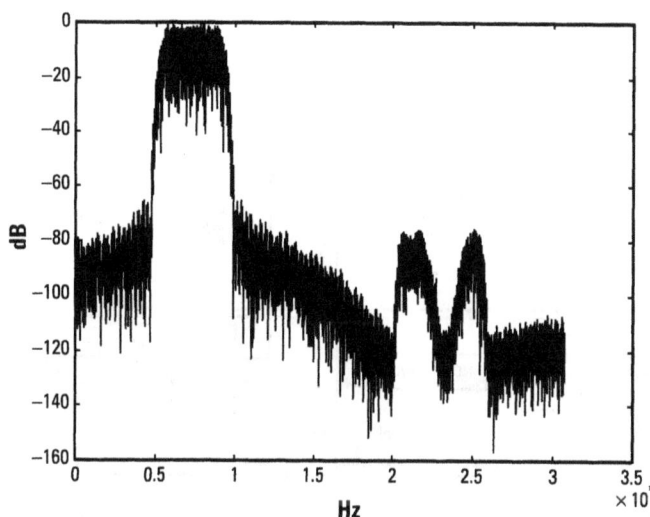

Figure 5.16 ISL5217 UMTS analog spectrum. (*Source:* Intersil, 2001. Reprinted with permission.)

The full-band spectral response in Figure 5.16 indicates an SFDR of almost 80 dB.

5.5 DDCs

Two of the most recent 3G-capable digital downconverters from Intersil are detailed in this section.

5.5.1 ISL5216

The maximum output bandwidth for each channel of this quad downconverter is 1 MHz. Higher output bandwidths are achievable by cascading filters and polyphasing the outputs [10]. All four channels must be cascaded to achieve a bare minimum of UMTS performance for a single 2X oversampled output. For improved UMTS efficiency and better performance the ISL5416 is recommended over the ISL5216. This device is, however, acceptable for downconversion of narrowband channels (e.g., GSM and EDGE).

At the time of writing, the ISL5416 is a very new device and information is still being made available; the ISL5216 is included here to provide sufficient downconverter coverage. Figure 5.17 illustrates ISL5216.

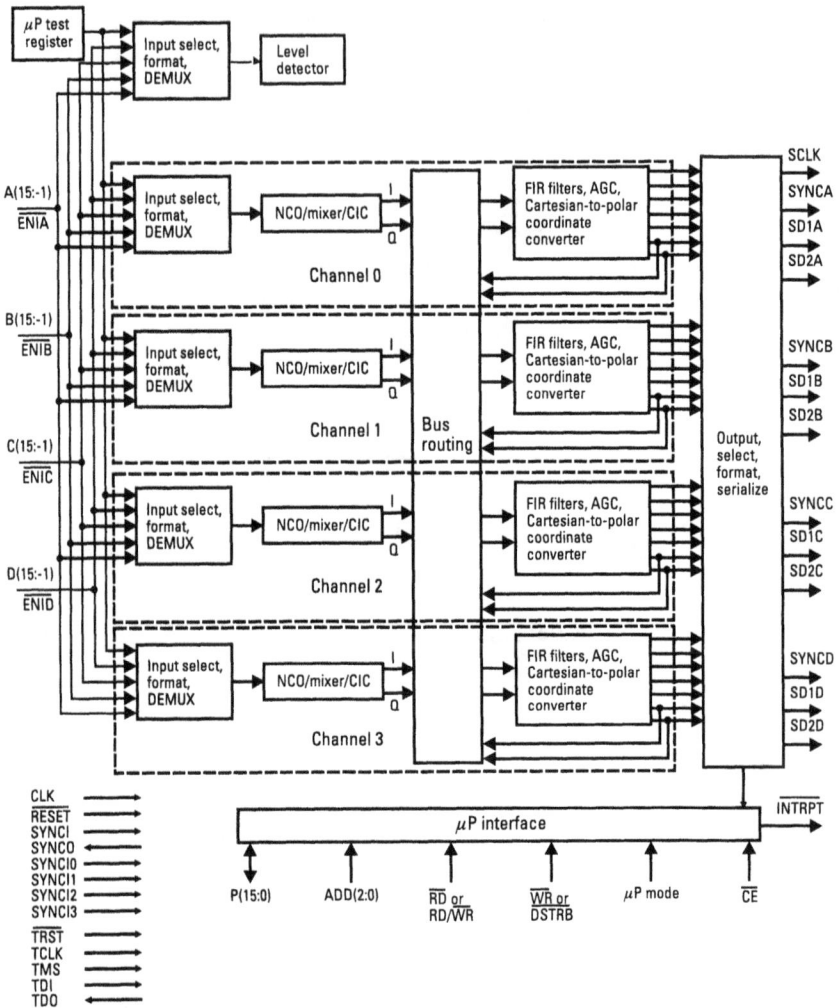

Figure 5.17 ISL5216 block diagram. (*Source:* Intersil, 2001. Reprinted with permission.)

5.5.1.1   CIC Filter

The CIC filter can be configured between one and five stages. Decimation is programmable from 4 to 65,536, or a maximum of 512 if a full five-stage filter is employed. A barrel shifter is included prior to the CIC filter to provide gain control and keep the overall gain between 0.5 and 1.0.

5.5.1.2   Filter Compute Engine

The FIR filters, AGC, Cartesian-to-polar coodinate converter processing block includes a flexible filter compute engine [11], which is programmed in a way similar to a microprocessor. It is a dual multiply accumulator data path with a microcoded FIR sequencer. The engine can implement a single FIR or a set of filters. The microcode can be a maximum of 32 sequences, where each sequence can be a filter or one of four flow instructions (i.e., Jump, Wait Load, Loop Counter, or NOP).

5.5.2   ISL5416

The ISL5416 is a four-channel wideband programmable 3G-capable downconverter from Intersil. It is illustrated in Figure 5.18.

5.5.2.1   Input Functions

The ISL5416 has four 17-bit inputs—A, B, C, and D—each with its own clock to allow for timing skew for cases where there are multiple ADCs feeding the downconverter. Inputs will accept a maximum sample rate of 80 MSPS, where the data format can be represented as 16-bit fixed point or 17-bit floating point. The range control and RF attenuator VGA control blocks are provided for RF gain control of the IF input to the ADC, which, in turn, feeds the downconverter. This range control function provides a very low latency way of avoiding clipping of the ADC output by reducing the input signal amplitude. The EOUT bus can be connected to four digital VGAs or digital RF attenuators using 4 bits each or to two control devices using 8 bits each. The input channel routing block provides switching to allow any input to be connected to any downconverter channel; this includes inputs being connected to multiple channels.

5.5.2.2   Mixing and Filtering

Each of the four channels contains an NCO/Mixer to perform the frequency downconversion to baseband followed by CIC filtering, channel filters, AGC, and a resampling filter.

5.5.2.3   Output Routing and Formatting

Output routing and formatting blocks are provided between the resampling filter block and the eight 8-bit parallel output buses. The output bus can be configured as eight 8-bit outputs, four 16-bit outputs, or two 32-bit outputs.

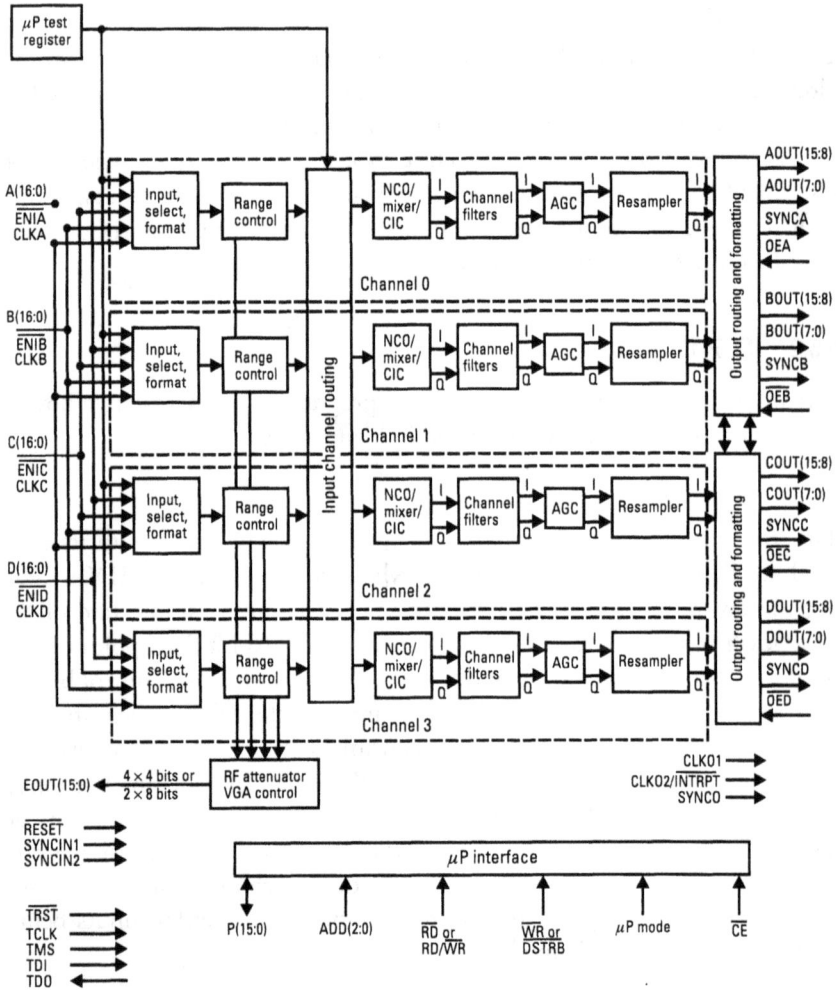

Figure 5.18 ISL5416 block diagram. (*Source:* Intersil, 2001. Reprinted with permission.)

The multiplexing system allows combinations of data to be sent to the output bus, including I and Q, magnitude, or gain data from each of the four channels. It is possible to format the I and Q channel data as either 4, 6, 8, 12, 16, 20, or 24 bits, and up to 16 bits of gain data are available. I and Q data can be output separately or multiplexed if the channels are synchronized.

The ability of the multiplexing system to send many channels of data out on 8 or 16 pins reduces the pin count of the ISL5416 with a commensurate decrease in circuit board real estate and complexity. This design also offers the possibility of parallel to serial conversion for low-cost transmission of data by high-speed serial protocols (e.g., LVDS).

5.5.2.4 Performance

The ISL5416 is specified as targeting the 3G applications of WCDMA/ UMTS and CDMA2000. It accepts a maximum input sample rate of 80 MSPS, and each of the four data channels supports a 5-MHz channel bandwidth with full filtering. The 32-bit programmable NCO has an SFDR of greater than 100 dB.

Apart from the mixing of unwanted components with the NCO's spurious responses, the other consideration in downconverter performance is the overall filter response. The quality of the filter response can be measured by the size of the responses in the stopband. These stop-band responses will cause aliases that fold back into the passband and reduce the signal to noise ratio. Since the DDC and DUC devices detailed in this chapter are extremely programmable, their absolute performance is tightly coupled to the combination of filter programs used.

5.5.3 Cascading Digital Converters and Digital Frequency Converters

In Chapter 4 we provided performance figures for digital converters in isolation, but what performance is needed when they are cascaded with a digital frequency converter? The answer is that the designer must devise a system budget and allocate performance parameters to each function and an implementation margin. This must be followed by extensive simulation to confirm the budgets. Finally, the designer should construct a prototype of the digital processing chain and confirm the simulation by real-life measurement.

5.6 Conclusion

Just as digital converters have recently matured and increased their performance, significant investment has been poured into DDCs and DUCs to bring them to the point where they can now be incorporated in wideband 3G SDRs.

We continue our path along the signal processing chain in Chapter 6 and cover options for baseband processing. Software figures even more prominently in this chapter, as we look at DSPs and other high-performance signal processing platforms.

References

[1] Chester, D., "Digital IF Filter Technology for 3G Systems: An Introduction," *IEEE Communications Magazine*, February 1999.

[2] Donadio, M., "CIC Filter Introduction," http://www.dspguru.com, July 18, 2000.

[3] Hogenauer, E. B., "An Economical Class of Digital Filters for Decimation and Interpolation," *IEEE Transactions on Acoustics Speech and Signal Processing*, Vol. ASSP-29, No. 2, 1981, pp. 155–162.

[4] Ifeachor, E., and B. Jervis, *Digital Signal Processing: A Practical Approach*, Reading, MA: Addison-Wesley, 1993.

[5] Crochiere, R. E., and L. R. Rabiner, "Optimum FIR Digital Filter Implementations for Decimation, Interpolation, and Narrowband Filtering," *IEEE Transactions on Acoustics Speech and Signal Processing*, 1975.

[6] Crochiere, R. E., and L. R. Rabiner, "Further Considerations in the Design of Decimators and Interpolators," *IEEE Transactions on Acoustics Speech and Signal Processing*, 1976.

[7] Crochiere, R. E., and L. R. Rabiner, "A Program for Multistage Decimation, Interpolation, and Narrowband Filtering," *IEEE Programs for DSP*, 1979.

[8] Crochiere, R. E., and L. R. Rabiner, "Interpolation and Decimation of Digital Signals—A Tutorial Review," *IEEE Proceedings*, 1981.

[9] McKinney, D., "ISL5217 Cellular Applications," Intersil Application Note AN9911, May 2001.

[10] Algiere, A., and D. Radic, "Use of HSP50216 and ISL5216 QPDC in Wideband Applications—UMTS," Intersil Application Note AN9927, April 2001.

[11] Intersil, "HSP50216 Data Sheet," April 2001.

6

Signal Processing Hardware Components

In this chapter we will review several software programmable silicon chip technologies; candidates were selected based on their potential to perform baseband (symbol-rate and chip-rate) signal processing in a software radio. The hardware architecture of each chip is provided, along with information on the tools and compilers that can be used during the software development process. The review includes figures for theoretical signal processing capacity and an estimate of the number of algorithm instances that may be able to fit on a device. The processing requirements for a 3G UMTS receiver and transmitter are used as the benchmark for the capacity calculations.

6.1 Introduction

Digital signal processors (DSPs) are the largest and most popular elements used today to perform hard, real-time signal processing. However, the increased need for radio frequency (RF) bandwidth introduced by IS-95 CDMA and exacerbated by UMTS WCDMA has resulted in the recent development of a new class of reconfigurable processors. These devices increase the degree of processing parallelism to the extent that they can efficiently tackle CDMA chip-rate processing. Field programmable gate arrays (FPGAs) have been popular for prototyping systems prior to implementation in ASIC, and now manufacturers are promoting them as alternatives to DSPs.

6.2   SDR Requirements for Processing Power

TDMA systems (e.g., IS-136 and GSM) multiplex approximately three to eight users (voice or data call) on a carrier and use RF bandwidths in the region of 30–200 kHz. CDMA systems (e.g., IS-95, CDMA2000, and UMTS) increase the multiplexing rate to approximately 50 to 200 users on a carrier; they use spread spectrum, where the RF channel bandwidth is increased to between 1.25 and 5 MHz.

CDMA uses two complex techniques to reduce the effects of multipath fading and improve system performance. At the network level soft handover [1] is used, and at the air interface level a type of receiver known as the Rake is implemented. In contrast, TDMA systems employ hard handover (break before make) when a call is transferred from one base station to the next; the call can be lost if significant fading occurs during the critical message signaling process. CDMA's soft handover (make before break) technique can reduce lost calls; during migration across cell boundaries the mobile terminal is always connected to two or more base stations. The call is only released from the preceding base station if the call has been fully established by the target base station. This reduces the probability of a loss of communication during severe fading on one (preceding or target) of the links.

Price and Green [2] first described the Rake receiver when they presented an implementation for use in the HF band. As the idea of a garden rake suggests, the receiver has several prongs; these are commonly referred to as "fingers" by the mobile communications industry. Each finger is a receiver in its own right, and each is able to lock onto a discrete multipath signal so long as the delays for each path are sufficiently separated in time. The input to each finger is a baseband chip-rate spread and modulated signal; the finger despreads the signal by correlating it with a delayed copy of the spreading waveform applied by the transmitter. Each finger outputs modulated symbol rate signals, which are combined to increase signal to noise ratio. Typically a base station Rake receiver will employ four fingers, and a mobile terminal Rake will use three.

For more TDMA and CDMA air interface details see Chapter 8, and for an in-depth coverage of the Rake concept, please see [3].

CDMA trades improved system performance for a significant increase in both system complexity (particularly in the terminal and BTS) and signal processing power. The UMTS receiver is well recognized as the most signal processing intensive receiver design used thus far for cellular mobile communications. It is easy to see why; the Rake is effectively several receivers in one and the higher chip rate used in UMTS (3.84 Mcps) increases the data rate into the receiver input. Also, because UMTS base stations are not synchronized (IS-95 and CDMA2000 are synchronous systems), there are more operations to be performed during path searching and handover.

Designers of multiple air interface (e.g., UMTS, CDMA2000, GSM, TDMA) software defined platforms who are in the process of dimensioning the signal processing system can use the UMTS receiver and transmitter as the worst case for consumption of signal processing resources. The information in Table 6.1 estimates the signal processing load for a single UMTS receiver and transmitter considering a 384-Kbps data channel (sourced and modified from [4]).

Table 6.1 has been expanded from [4] by estimating that a UMTS transmitter requires approximately 33% of the processing power of a UMTS receiver.

In Chapter 5 we covered channelization (digital frequency up- and downconversion) and proposed that this function be partitioned to digital upconverters and downconverters; the 3,000 MIPS figure in Table 6.1 is in general agreement with the estimate provided in Table 2.2 (i.e., 4,340

Table 6.1
UMTS Signal Processing Load

Function	Partition	Tx/Rx	MIPS
Channelization	IF	Rx	3,000
Path searcher	Chip rate	Rx	1,500
Access detection	Chip rate	Rx	650
Rake receiver	Chip rate	Rx	650
Maximal ratio combining	Chip rate	Rx	24
Channel estimation	Symbol rate	Rx	12
AGC, AFC	Symbol rate	Rx	10
Deinterleaving, rate matching	Symbol rate	Rx	12
Turbo decoding	Symbol rate	Rx	52
Channelization	IF	Tx	3,000
Transmitter	Chip rate	Tx	900
Interleaving	Symbol rate	Tx	12
Turbo encoding	Symbol rate	Tx	15
		Total:	9,837

MIPS). The four major chip-rate functions are path searcher, Rake receiver (including maximal ratio combining), access detection, and transmitter. Table 6.1 also shows that the symbol-rate functions consume significantly fewer signal processing resources than the chip-rate functions.

The actual total requirement for chip-rate processing will vary considerably from the 3,724 MIPS figure (Path Search = 1,500 + Access Detect = 650 + Rake = 674 + Tx = 900) in Table 6.1. Processing power will depend upon system factors such as spreading rate, channel type, and cell range and will be heavily influenced by implementation decisions, including probability of access detection, algorithm revisit rates, and oversampling rate. UMTS uses variable rate spreading, where lower data rate channel types (e.g., voice) use higher spreading factors and therefore require more correlations.

The path search function is the most demanding of processor cycles. The required number of MIPS is directly dependent upon the oversampling rate, number of users, cell size, correlation size, and algorithm revisit rate. Path searching is performed by cross-correlating the incoming data sequence with part of the spreading sequence used to transmit the waveform. For periodic continuous waveforms, $x(t)$ and $y(t)$, with period T, the cross-correlation function is defined as:

$$R_{xy}(\tau) = \frac{1}{T} \int_{0}^{T} x(t) \Box y(t - \tau) dt \qquad (6.1)$$

Equation (6.1) shows that cross-correlation is multiplication intensive for continuous waveforms. Because a CDMA despreading sequence is a series of positive and negative ones (e.g., +1, −1, −1, +1, ... +1, and so on), cross-correlation merely becomes a series of additions and subtractions [i.e., the functions found in an arithmetic logic unit (ALU)]. When choosing a processing device to perform path searching, it is more important to concentrate on ALU performance than the number of dedicated multipliers and the MMACS figure.

Now consider a 4X oversampled UMTS 3.84-Mcps signal, where the correlation sequence length is 128. Baseband data will enter the path searcher at $4 \times 3.84 \times 10^6$ = 15.36 Msps and require $15.36 \times 10^6 \times 128$ correlations per second per search chip. For a cell size of 500m the path searcher will have to search over the return path plus an allowance for delay spread, assuming a worst-case spread of 50%; this is equivalent to 1,500m or 19 chips. Assuming the mobile is moving within normal limits, path delay will change relatively (to the frame rate) slowly and therefore the search algorithm can be conducted at a less than 100% duty cycle, (e.g., 5%). For this case the path searcher will be performing approximately $4 \times 3.84 \times 128 \times 19 \times 0.05 = 1,868 \times 10^6$ correlations per second per user.

For a 4X oversampled CDMA2000-1X 1.2288-Mcps signal using a similar set of conditions as for the UMTS case, the number of correlations will reduce by the ratio of the two chip rates (i.e., 3.84/1.2288). The CDMA2000 path searcher will be performing approximately 467×10^6 correlations per second per user.

6.3 DSPs

Moore's Law is now widely accepted as part of the microprocessor vernacular; it relates to the compound doubling of computer processing power that has been observed to occur approximately every 18 months or so. These advances have been possible due to improving designs and chemical processes that have shrunk the size of transistors (halving every 18 months), increased clocking speeds (doubling every 18 months), and reduced power consumption. There are theories that this exponential improvement will hit a ceiling due to physical effects such as unrealizable clock speed or gamma radiation entering the silicon chip and causing false logic state changes. So far this has proved not to be the case, and recent reports indicate that Moore's Law will be able to support another decade of improvement at least.

Gordon Moore reported this trend in 1965 and since then it has been mostly associated with general-purpose microprocessors. DSPs have followed a similar trend, as illustrated in Figure 6.1, where million multiply and accumulations per second (MMACS) are plotted over time for the Texas Instruments range of DSPs. The MAC is a good overall performance indication for the DSP as applied to software radio, because many radio functions are multiply and accumulation intensive.

Therefore, on average, every 18 months we can expect a doubling in processing power for the same volume, power consumption, and cost. This equates to an order of magnitude (or 10X) improvement every six to seven years. It is this exponential improvement that has leveraged the software radio from the university laboratory into the commercial marketplace.

There is also an argument [5, 6] that the actual requirement for processing power needed by the wireless industry is exceeding Moore's Law. Investigations into the increase in algorithmic complexity experienced during the transitions from 1G to 3G suggest that the actual need for processing

Software Defined Radio for 3G

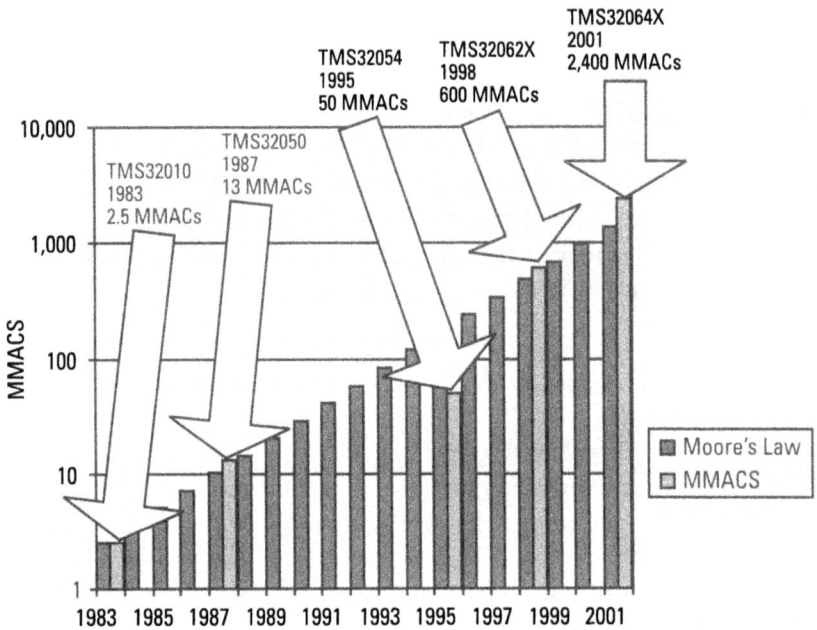

Figure 6.1   Moore's Law and DSP MMACS versus time.

power is increasing by an order of magnitude every 4 years. The proponents
of this idea suggest that new architectural technologies are required to keep
pace, and the industry cannot rely on faster versions of traditional processing
architectures (e.g., Super Harvard and others) to solve the need for more signal processing power.

6.3.1 DSP Devices

6.3.1.1 Texas Instruments, C64X DSP

The C6000 family of DSPs started life in 1997 with the C62X and C67X
cores; the C6000 family uses an advanced very long instruction word
(VLIW) architecture. The architecture contains multiple execution units
running in parallel, which allows them to perform multiple instructions in a
single clock cycle. The latest member of the C6000 family is the C64X; its
architecture can perform twice as many multiply operations as the C62X,
and the first available devices can be clocked at twice the speed. As shown in
Figure 6.1, the C64X kept up with Moore's Law and theoretically quadru-
pled the number of MMACS when compared with the C62X. When dimen-

sioning a signal processing system, theoretical (assumes every cycle is usefully employed) MMACS values should be used judiciously, since the realizable performance will depend upon many factors, including the subject algorithms and compiler efficiency.

The C64X has a clock speed road map from 600 to 1,100 MHz; this will provide a further 83% increase in signal processing power.

The C64X is shown in Figure 6.2.

The C64X core consists of eight functional units, two register files, and two data paths. As with the C62X, two of the eight functional units are multipliers; however, each multiplier has been enhanced to allow two 16-bit or four 8-bit multiplies every clock cycle. Eight-bit data are common for image processing but not so popular for wireless applications, as indicated in previous chapters. Software radio generally uses 12–16-bit data to meet 3G per-

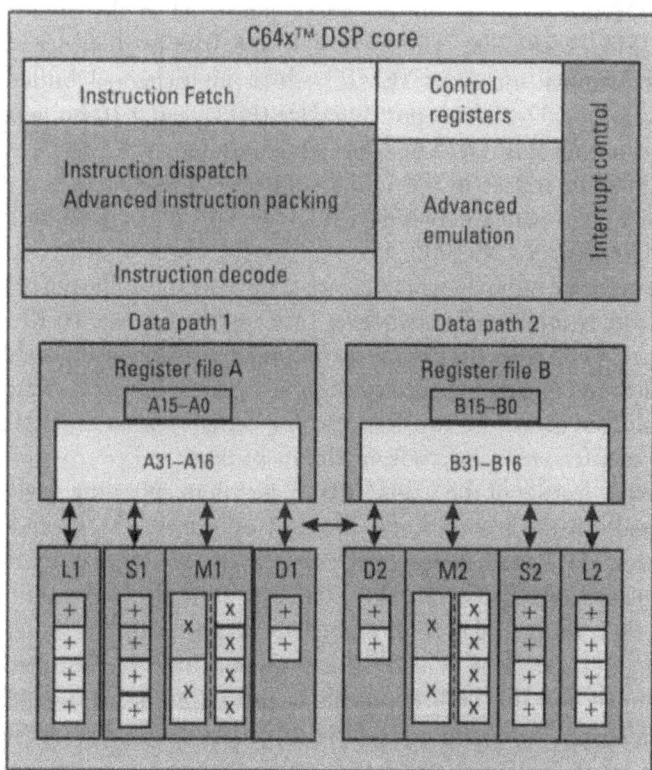

Figure 6.2   TMS320C64 core block diagram. (*Source:* Texas Instruments, 2001. Reprinted with permission.)

formance parameters. The 16-bit extensions built into the multiply functional units are also present in the six other functional units. These include dual 16-bit addition/subtraction, compare, shift, min/max, and absolute value operations.

Packed 8-bit and 16-bit data types are used by the code-generation tools to take advantage of the extensions. By doubling the registers in the register file and doubling the width of the data path, as well as utilizing instruction packing, the C6000 compiler has been made more efficient by having less restrictions placed on it by the chip's architecture.

The approach of evolving the C62X architecture by increasing clock speed and expanding the architecture (instead of designing a completely new architecture) allows the C64X to be object code compatible with the C62X.

The C64X device is available in three versions: C6414, C6415, and C6416. Each device uses the same core and memory interfaces to level 1 and 2 cache. Devices differ by the interfaces connected to the direct memory access (DMA) controller. The C6414 is the base-level device with two enhanced memory interfaces (EMIF), three multichannel buffered serial ports (McBSP), a 32-bit host port interface (HPI), and a 16-bit general-purpose input/output (GPIO). The C6415 expands upon the C6414 by providing additional interfaces to the DMA controller. These include a choice of Peripheral Component Interconnect (PCI) or HPI plus GPIO and a choice between Utopia 2 or a McBSP.

The software radio designer's choice is the C6416, as illustrated in Figure 6.3. The core is connected to two level 1 caches; the first is a 16-Kbyte direct mapped program cache (L1P) and the second a 16-Kbyte data cache (L1D). Each of the level 1 caches is connected to a large 1-MB level 2 cache, unified for program and data. This two-level caching design automates off-chip to on-chip data transfers, reducing code development time.

A major feature of the C6416 that makes it an appealing device for 3G software radio is the addition of a Viterbi Coprocessor (VCP) and a Turbo Coprocessor (TCP). In a way, the coprocessors are similar to the ASSP devices covered in Chapter 5. The VCP and TCP are application specific; however, they are provided with enough programmability to be applied to the leading 3G standards. The VCP can support over five hundred 8-Kbps voice channels and the TCP is capable of processing up to thirty-five 384-Kbps data channels. Viterbi convolutional codecs are used on speech traffic to correct for transmission errors and can be found in 2G and 3G systems. The human brain is reasonably tolerable of errors in speech; however, data transmission is significantly more demanding of link quality. 3G data ser-

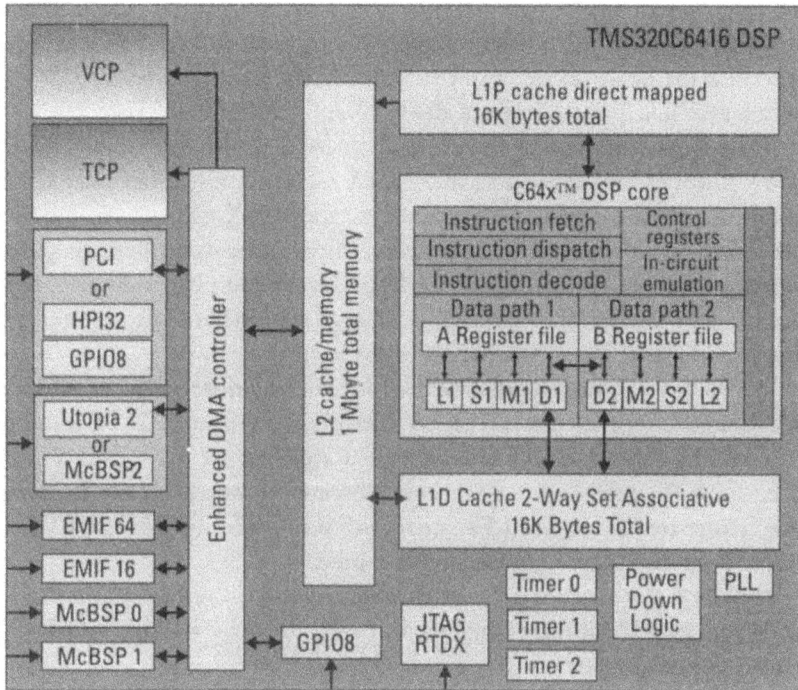

Figure 6.3 TMS320C6416 DSP block diagram. (*Source:* Texas Instruments, 2001. Reprinted with permission.)

vices use the much stronger error correcting capability offered by the Turbo Codec. The TCP and VCP perform decoding only; encoding is significantly less complicated and can be performed on the DSP core. The dedicated processing capability of the coprocessors frees up processing power on the DSP core and allows more instances of core functions to be implemented on the C6416.

Some measures of the C6416's improved performance include a nearly 15 times increase in symbol rate performance over the TMS320C6203. For a 16-state GSM Viterbi decoder the C64X core, when clocked at 600 MHz, can achieve a 5.4 times performance improvement over the C62x, when clocked at 300 MHz.

6.3.1.2 Texas Instruments C55X DSP

While the C6000 family of DSPs is targeted at infrastructure projects such as cellular mobile BTSs and Node Bs, the C5000 family has been primarily developed for power drain–sensitive portable devices such as cellular mobile phones and wireless terminals. Texas Instruments has progressed the C54X road map to produce the latest device, a software source code–compatible C55X. This device is available in three variants: the C5502, C5509, and C5510.

The family is designed for a core dissipation as low as 0.05 mW/MIPS at 0.9V and performance up to 800 MIPS using a 400-MHz clock. Compared with the previous C54X generation, the C55X will deliver approximately five times the performance and dissipate one-sixth the core power dissipation. Part of the reason for the large decrease in power consumption, despite a higher clock rate, is the inclusion of sophisticated power management techniques that are transparent to the user.

Considering some popular cellular mobile algorithms, the C55X consumes less energy than the C54X by the following factors [7]: GSM Full Rate Vocoder 5.8X, GSM Extended Full Rate Vocoder 6.2X, and Viterbi Codec 6.7X. In terms of an MIPS performance comparison, the improvement factors are 5X for GSM Extended Full Rate Vocoder, 7.5X for Convolutional Encoder, and 5.25X for Viterbi Codec.

Figure 6.4 illustrates a block diagram of the C5509 DSP; the C55X core is common to all three versions, but they differ by the peripherals and memory configurations.

The DSP supports an internal bus structure with one program bus, three data read buses, two data write buses, and additional buses for peripheral and DMA use. This architecture allows up to three data reads and two data writes in a single cycle. The DMA controller can act independently of the core CPU and perform in parallel up to two data transfers per cycle. There are two MAC units, each capable of performing addition, subtraction, and 17-bit multiply (fractional or integer) operations in a single clock cycle. The core is connected to a 32 Kword ROM, a 32 Kword Dual Access RAM, and a 32 Kword Single Access RAM.

The C55X has many of the same peripherals as the C64X with a few interesting additions. A 10-bit ADC has been included on the C5509 device; this could be utilized in a cellular mobile terminal using a completely analog RF front end. There is also a USB port for direct connection to other popular communications devices.

6.3.1.3 Analog Devices' TigerSHARC DSP

The ADSP-TS101S TigerSHARC DSP is a completely new design of DSP from Analog Devices aimed at high-powered infrastructure projects such as 3G BTSs and Node Bs. When clocked at 180 MHz, the device is capable of

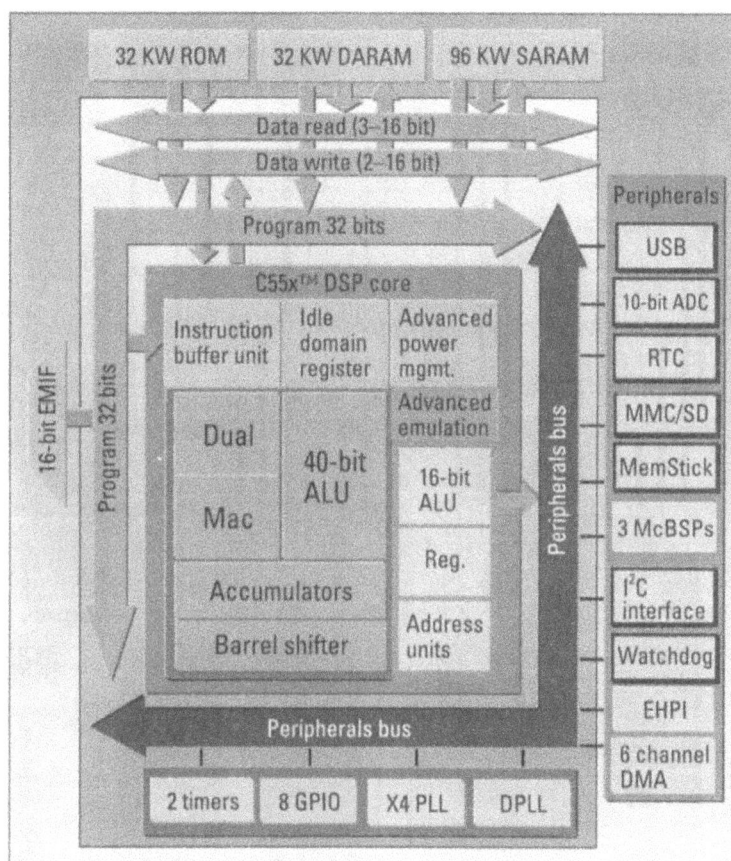

Figure 6.4 TMS320C5509 DSP block diagram. (*Source:* Texas Instruments, 2001. Reprinted with permission.)

a peak of six instructions per cycle, or 1,080 MIPS [8]. In terms if 40-bit MACS, the ADSP-TS101S is capable of a peak of 1,440 MMACS and claims the position of world's highest-performance floating-point DSP [9]. The chip architecture is described as static superscalar and allows the processing of 1-, 8-, 16-, and 32-bit fixed-point and floating-point data types on the one device. Static superscalar is viewed as a combination of reduced instruction set computing (RISC), VLIW, and DSP functionality. A superscalar processor is one that is capable of sustaining an instruction-execution rate of more than one instruction per clock cycle [10]. The TigerSHARC DSP includes the key superscalar design features of a load/store architecture,

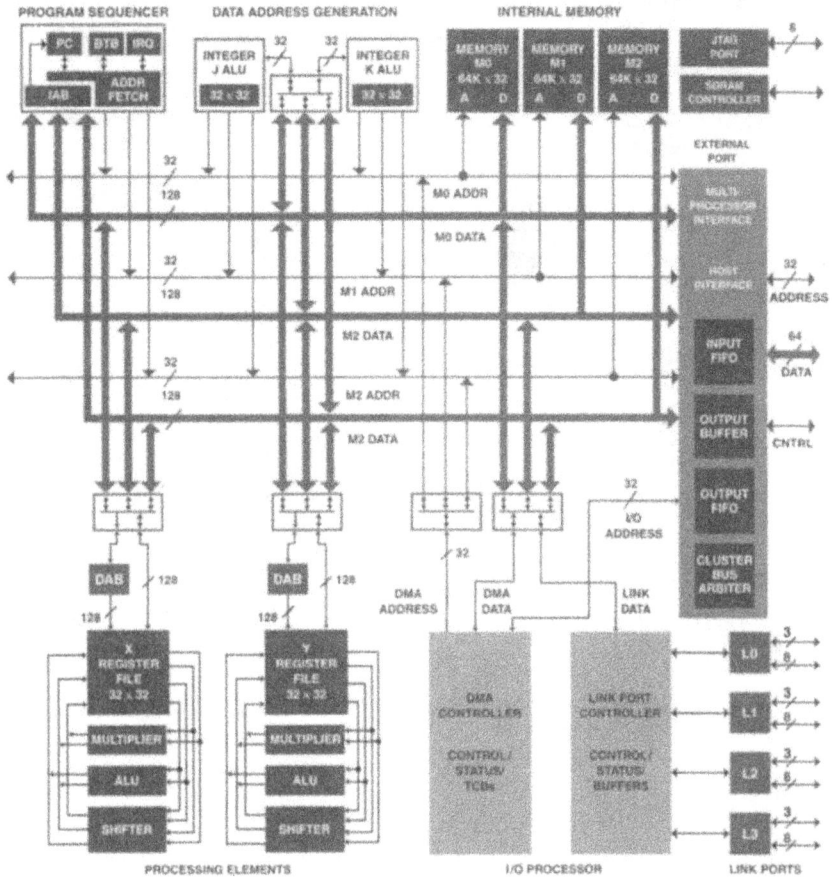

Figure 6.5 ADSP-TS101S block diagram. (*Source:* Analog Devices, 2001. Reprinted with permission.)

branch prediction, and a large interlocked register file. The TigerSHARC
DSP is shown in Figure 6.5.

The TigerSHARC DSP's two computation blocks (X and Y) each contain a multiplier, arithmetic logic unit (ALU), and a 64-bit shifter. With this
configuration of functions it is possible in a single clock cycle to execute
eight 16-bit MACS, two 32-bit MACS, or two 16-bit complex MACS.
Clocking at 180 MHz [8], these per cycle figures convert to twenty-eight
hundred eighty 8-bit MMACS, fourteen hundred forty 16-bit MMACS,
three hundred sixty 32-bit MMACS, or three hundred sixty 16-bit complex
MMACS.

Some 16-bit benchmark algorithm performance figures include a 256-point complex FFT in 7.3 µs and a 50-tap FIR filter in 48 µs. Completing a 3G turbo decode on a 384-Kbps data channel will consume 51 MIPS, a Viterbi decode on a 12.2-Kbps voice channel will take 0.86 MIPS, and a 3.84-Mcps UMTS complex correlation using a spreading factor of 256 will consume 0.27 MIPS. These benchmarks can be converted to algorithm instances as follows: 3 Turbo decoders, 209 Viterbi decoders, and 667 complex correlations.

There are three 64-Kbyte blocks of 32-bit internal memory; each block can store a variety of data types up to "Quad" (four 32-bit words). The DSP provides a FIFO buffered 64-bit external port, a 14-channel DMA controller, and four bidirectional link ports capable of 720 MB per second transfer rate. Many of Analog Devices' DSPs include this link port arrangement, which allows four other DSPs to be connected without any additional external glue logic. This has the advantage of reducing the number of on-board integrated circuits when constructing a processing array of many DSPs.

6.3.1.4 Motorola's MSC8102

Motorola and Lucent's Micro-Electronics Group (now Agere Systems) have jointly developed a DSP core called the StarCore SC140. As a core this can be incorporated with other functions into an ASIC to produce a system on a chip (SoC). Motorola has proceeded down this track and released several DSPs that use the SC140 core, including the MSC8102.

The MSC8102 incorporates four SC140 cores and 1.5 MB of on-chip memory; each core can be clocked at 300 MHz. Each SC140 core is capable of a theoretical 1,200 MMACS (4 MACS per clock cycle), and each is connected to a 300-MHz Enhanced Filter Coprocessor (EFCOP). The additional filter hardware boosts the overall signal processing capacity of the MSC8102 to 6,000 MMACS [11]. The EFCOPs are designed for wireline applications (e.g., echo cancellation); therefore, the extra 1,200 MMACS they provide may not be directly useful for software radio applications. In terms of MIPS, each core is capable of up to six parallel instructions (four arithmetic + two moves) per clock cycle; this equates to 1,800 MIPS per core or 7,200 MIPS per MSC8102. The six instructions are quoted as being equivalent to ten RISC instructions [11]. RISC MIPS may not be readily comparable with the MIPS values provided by the C64X (VLIW architecture) and TigerSHARC DSP (mix of VLIW, RISC, and DSP).

The memory architecture for each core is segmented as 224 KB of dedicated level 1 SRAM and 16 KB of instruction cache. The MSC8102 has

476 KB of level 2 memory, which is shared by the four processor cores. In terms of I/O the chip has a standard PowerPC bus interface, a 32- or 64-bit direct slave interface, and four independent time division multiplex (TDM) interfaces.

6.3.2 DSP Performance Summary

Using the manufacturer's stated performance figures it is possible to produce a DSP performance comparison. To complete this comparison we need to quantify the power of the C64X coprocessors. Comparing the TigerSHARC DSP with the TCP and VCP inside a C6416 600-MHz device, the TCP can process 35 channels or 1,167% (35/3) of a TigerSHARC DSP's complete capacity, and the VCP's 500 channels are equivalent to 239% (500/209) of a TigerSHARC DSP. The TCP can be translated to 1,167% of 1,080 MIPS or 12,604 MIPS, and the VCP can be translated to 239% of 1,080 MIPS or 2,581 MIPS. The figures for the 1,100-MHz C6416 coprocessors are increased proportionally to the clock frequency change. The final comparison for DSPs biased toward infrastructure projects (e.g., BTS) is provided in Table 6.2.

Assuming that the EFCOP units on the MSC8102 are not useful for SDR, and Turbo and Viterbi decoding are required, the C6416 1,100-MHz device provides the largest peak performance with 36,639 MIPS (8,800 + 23,107 + 4,732). This comparison can be used as an input to the DSP selection process; however, any serious tradeoff study will consider many more factors, such as compiler efficiency (see Section 6.2), I/O capa-

Table 6.2
DSP Performance Summary

	DSP			
	C6416, 600 MHz	C6416, 1100 MHz	TigerSHARC, 180 MHz	MSC8102, 300 MHz
8-bit MMACS	4,800	8,800	2,880	—
16-bit MMACS	2,400	4,400	1,440	4,800
MIPS	4,800	8,800	1,080	7,200
Coprocessors	TCP (12,604 MIPS) + VCP (2,581 MIPS)	TCP (23,107 MIPS) + VCP (4,732 MIPS)	None	EFCOP (1,200 MMACS)

bility, cache structure, instruction set functionality, power dissipation, code size, and legacy issues.

6.3.3   DSP Compilers

When the first single-digit MMAC DSPs came on the scene in the early 1980s, the only way they could be programmed efficiently was by using assembly code. This method required the software developer to learn a new instruction set and language for each brand of DSP, and often a relearning exercise was required for each new device generation of the same brand.

Since those early days many tens of thousands of engineering days have been invested in developing compilers that accept a high-level language (e.g., C or C++) input. These compilers are now producing executable code with performance close to that achieved by hand-crafted assembly code. High-level languages have the following advantages: Software can be developed more efficiently, they are more easily reviewed by peers, and the source code is more portable. Increasingly, time to market is the most important factor in the product development cycle. It is now possible that the combination of an excellent compiler and its matching DSP can outweigh a competing combination of a theoretically more powerful DSP (MIPS or MMACS) and its matching compiler of only average performance.

6.3.3.1   Code Composer Studio

Code Composer Studio is Texas Instruments' compiler and integrated development environment (IDE) for the C5000 and C6000 range of DSPs; it allows developers to use either C or C++ high-level languages. As an IDE it includes an integrated compiler and debugger interface that does not require the user to switch between applications. It supports multisite connectivity (for geographically dispersed development teams) for single or multiple processor systems. Systems can be of single processor type or mixed (i.e., both C5000 and C6000). Multiplatform (UNIX and PC) development is supported by the ability to handle external make-files.

Emulation capabilities are provided in a real-time data exchange facility that can support between 2 and 50 C5000 and C6000 devices. Real-time debugging is also supported by a DSP/BIOS facility.

The Code Composer compiler provides four levels (o0, o1, o2, and o3) of optimization. The prime job of the optimizer is to reduce overall execution time. Simple optimization can include techniques such as unraveling loops to reduce the number of branch instructions. More complex approaches involve

Table 6.3
Code Composer Studio Efficiency

Algorithm	Use	Assembly Cycles	C Cycles	Percent Efficiency
Codebook Search	CSELP vocoder	977	976	102
10-Tap FIR Filter	VSELP Vocoder	238	280	85
16-Tap IIR Filter	Filter	43	38	113

the optimizer automatically analyzing sections of the C/C++ code for recognizable patterns that can be most efficiently mapped onto the architecture using the DSP's instruction set. The optimizer is benchmarked against a large array of functions typically used in wireless applications; this ensures that compiler upgrades always improve performance.

Examples of the efficiency of Code Composer when compared with hand-crafted assembly are provided in Table 6.3 [12].

6.3.3.2 Analog Devices' CROSSCORE

The IDE provided for the TigerSHARC DSP is called CROSSCORE, which includes VisualDSP++; this software development tool also supports the more general-purpose SHARC and Blackfin range of Analog Devices' DSPs.

DSP software can be developed using either the C or C++ programming languages, and the debugger supports multiprocessor configurations. Version 2.0 of the tool has been designed specifically to work with the ADSP-TS101-S DSP and includes a lightweight kernel (operating system). The VisualDSP++ kernel allows the user to create and delete threads, as well as work with semaphores, events, and other real-time embedded programming features.

6.3.3.3 CodeWarrior

CodeWarrior is a visual development environment developed by Metrowerks [13] for the StarCore core and the Motorola StarCore–based DSPs. The CodeWarrior compiler accepts the C programming language and was designed to take advantage of the StarCore architecture. It includes more than 100 optimizations aimed at increasing performance and reducing code size. Apart from an integrated compiler and debugger, the tool includes profiling capabilities and a cycle accurate instruction set simulator.

6.4   Reconfigurable Processors

Considering the raw DSP performance figures provided in Table 6.2 and the chip rate processing requirements in Table 6.1, one can estimate the number of users (instances) of chip-rate processing that can fit onto the various DSPs; this is detailed in Table 6.4.

Considering that a CDMA ASIC will support a full carrier's load of chip- and symbol-rate processing, the problems associated with using tens and possibly hundreds of DSPs to perform only chip-rate processing are obvious. Such a choice will trade off "software defined" for increased cost and complexity. This potential dilemma has led to the development of a new class of reconfigurable processors; these are covered in Sections 6.4.1 and 6.4.2.

6.4.1   Chameleon Reconfigurable Communications Processor (RCP)

A lower parts count solution for chip-rate processing may be solved by replacing the DSP with a higher MIPS rated field programmable gate array. An FPGA's function is, however, static after boot-up; to change its functionality it must be taken off-line and programmed by a binary bit stream, usually taking fractions of a second. Therefore, all the algorithms to be run on an FPGA must be preloaded.

For systems where relatively large latency can be tolerated, greater FPGA efficiency can be achieved by implementing functions in pipelined

Table 6.4
DSP Chip-Rate Instances

	DSP			
	C64X, 600 MHz	C64X, 1100 MHz	TigerSHARC, 180 MHz	MSC8102, 300 MHz
MIPS	4,800	8,800	1,080	7,200
UMTS Chip-Rate Load (MIPS)	3,724	3,724	3,724	3,724
CDMA2000 Chip-Rate Load (MIPS)	931	931	931	931
Number of UMTS Users	1	2	<1	1
Number of CDMA2000 Users	5	9	1	7

mode. Pipelining allows serial algorithmic blocks to be concurrently active, with each block processing a different frame of data. This technique improves efficiency by utilizing more of the device for a higher percentage of the time. Latency of more than a frame in the air interface of a mobile telecommunications system is at best very undesirable and at worst may result in nonconformance to the standard. In mobile systems it is usual that all the data from the previous frame must be processed through several serial functions prior to the end of the current frame. Avoiding pipelining in order to meet latency requirements may result in an unacceptably low efficiency, since only a small fraction of the chip will be in use at any given instant.

The RCP is a new class of device specially developed to provide a very large signal processing capability and is well suited for wireless communications algorithms, including chip-rate processing. The design of this new processor avoids the static configuration problem of the FPGA by allowing the device's program to be completely changed from clock cycle to clock cycle. At any instant this allows the entire RCP to be dedicated to a single function and increase efficiency. The RCP's large signal processing capacity and relatively higher power dissipation make the device most suitable for infrastructure projects such as 3G base stations.

The RCP was developed by Chameleon Systems, Inc., a privately held start-up company formed in 1997 and is headquartered in San Jose, California.

6.4.1.1 Architecture

The major functional blocks of the RCP are illustrated in Figure 6.6.

The 32-bit embedded ARC processor is connected to the 32-bit reconfigurable processing fabric via a 128-bit RoadRunner bus. The fabric is the core signal processing engine and is discussed in more detail later in this section. The RoadRunner bus is capable of sustaining 1.6 GB/sec and connects to a PCI controller, configuration subsystem, DMA subsystem, and memory controller. Other on-chip buses include a 32-bit PCI bus and a 64-bit memory bus, which supports SDRAM, SSRAM, and flash memory.

The ARC processor delivers up to 100 MIPS at 100 MHz; it is capable of performing signal processing if needed but is normally used to control all the on-chip devices. The 64-bit memory controller is used for off-chip memory and can support 1 GB/sec transfer rates out to SSRAM and SDRAM. The DMA subsystem supports 16 DMA channels to transfer data between the ARC and the fabric. The configuration subsystem is a key piece of technology with a controller and two configuration planes, an active plane, and a

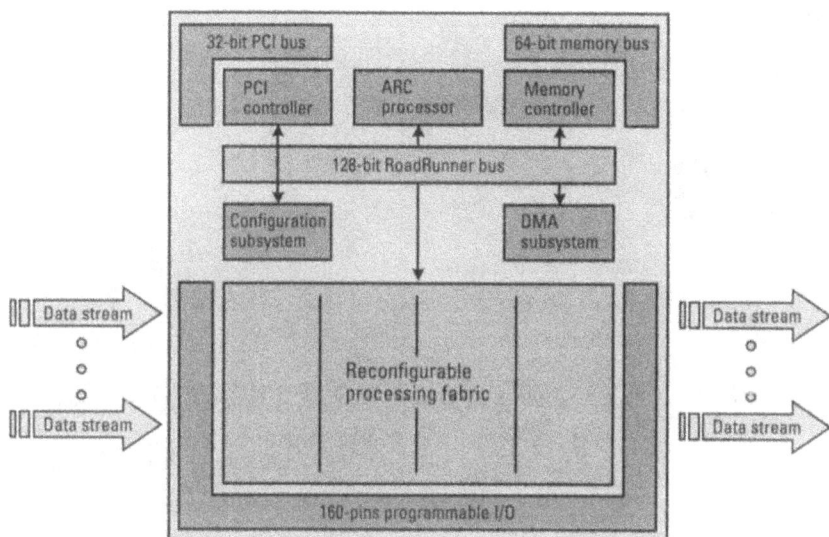

Figure 6.6 RCP major functional blocks. (*Source:* Chameleon, 2001. Reprinted with permission.)

background plane. The controller is an optimized DMA controller capable of transferring configuration data from off-chip memory to the background plane. Optimization and design allow the transfer to take place without disrupting the data flow and signal processing being conducted through the fabric.

The CS2112 reconfigurable processing fabric is divided into four slices, as illustrated in Figure 6.7.

Each slice can be independently programmed or configured, and each slice consists of three tiles. Figure 6.7 shows tile C or slice 0 exploded for clarity. Each tile has four local store memories (LSMs), seven data path units (DPUs), two multipliers, and a control and logic unit (CLU).

Two 16-bit by 24-bit single clock cycle multipliers are provided on each tile, with a total of 24 multipliers for the whole device. These multipliers alone provide 2,400 MMACS when a CS2112 is clocked at 100 MHz. The 40-bit signed result from a 16-bit by 24-bit product is truncated to 32 bits by removing the eight LSBs. In 16-bit by 16-bit mode the signed multiply produces a 32-bit result.

The four LSMs store one hundred twenty-eight 32-bit words; they are multiported to allow simultaneous read and write and can be aggregated to build a larger memory space. LSMs are directly accessible from the DMA subsystem and surrounding DPUs and multipliers.

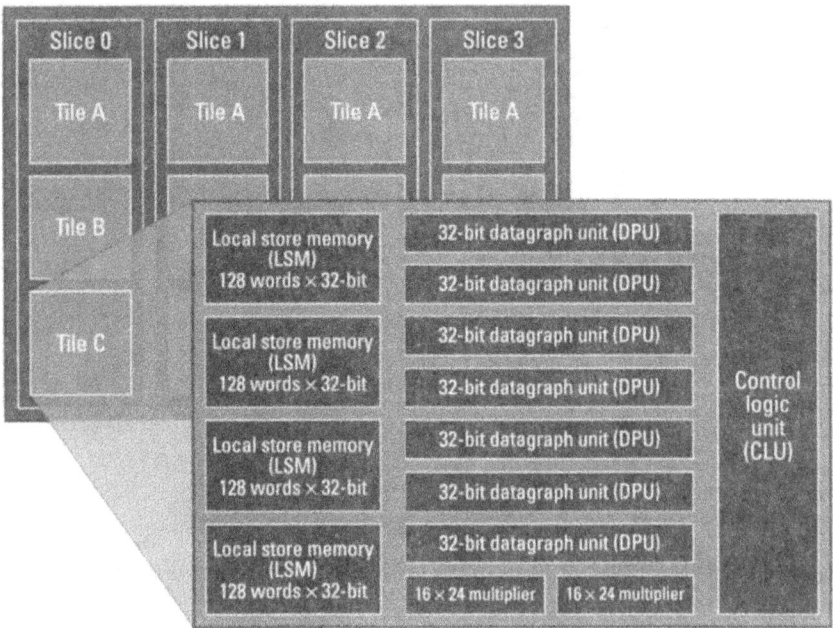

Figure 6.7   RCP fabric functional blocks. (*Source:* Chameleon, 2001. Reprinted with permission.)

The CLU is responsible for implementing finite state machine sequencing and conditional operation. It includes a programmable logic array (PLA), state bits, multiplexers, and control state memory (CSM). The state bits are used to select the DPU and multiplier instructions stored in the CSM. Each CSM stores eight DPU or multiplier configuration instructions. Based on the current state and PLA inputs, including flags passed from the DPUs, this finite state machine determines the next state and the next configuration instruction from among the eight stored instructions. The CLU also selects input and output routing and other configuration characteristics. The CLU is designed for conditional sequencing of operations depending on content.

The 32-bit DPU is shown in Figure 6.8.

Each tile has seven DPUs; these data processing modules directly support all C and Verilog operators. Thirty-two-bit operations are supported, and some 16-bit operations are implemented as dual 16-bit data streams (SIMD). Examples of supported operators include ADD, ADD16 (parallel 16-bit add), MAX, SADD (saturating add), and SUBCNT (32-bit subtract with control).

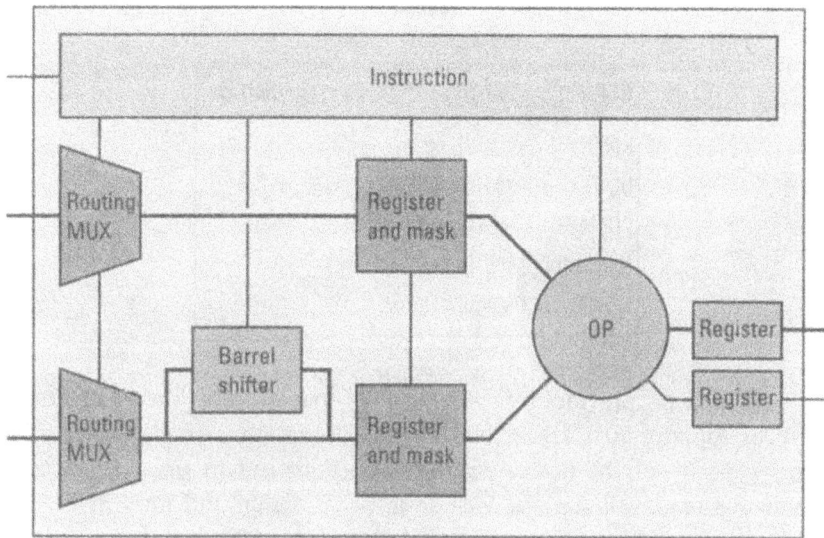

Figure 6.8 RCP data path unit functional blocks. (*Source:* Chameleon, 2001. Reprinted with permission.)

The fabric is 100% routable and timing is deterministic with a one-cycle delay for intraslice routes and two clock cycles for interslice routes. These characteristics are independent of fan-out. At the front of each DPU is a pair of 24:1 routing multiplexers (MUX); the inputs connect to local routes (± eight DPUs either side of the subject DPU), intraslice routes (all 21 DPUs in the slice), and interslice routes. The result of the MUX selection is an A and B operand; these feed the OP unit via a barrel shifter and register and mask in one leg and a register and mask in the other leg. The 32-bit real-time barrel shifter is capable of word swapping, byte swapping, and word duplication.

The DPU has an optional bit-shifted feedback mode for efficient linear feedback shift register (LFSR) implementations; this can be useful for the generation of pseudonoise sequences and spreading codes in CDMA.

Figure 6.6 shows 160 pins of programmable I/O (PIO); this is implemented as four banks of 40 pins, where each bank is capable of 0.4 GB/sec bandwidth or 1.6 GB total for all four slices. The PIO banks can be configured to provide interface and handshaking signals for memory, ADCs, DACs, FPGA, and so on.

Table 6.5 estimates the number of instances of chip-rate processing that may fit onto the RCP.

Table 6.5
DSP Chip-Rate Instances

RCP	CS2112
MIPS	19,200
UMTS Chip Rate Load (MIPS)	3,724
CDMA2000 Chip Rate Load (MIPS)	931
Number of UMTS Users	5
Number of CDMA2000 Users	20

Chameleon provides information [14] that supports the 2112's capability to support 50 CDMA2000 channels/users on a single device. The actual capacity of the device will be heavily dependent upon key system parameters (e.g., cell size), as well as software design and implementation efficiency. The simplistic use of MIPS (as per Table 6.5) does, however, provide a ballpark figure of the same order.

6.4.1.2   Tool Suite and Design Flow

Chameleon provides a tool suite (C-Side) and software design flow based around the UNIX environment for the RCP. A complete development suite includes simulators, compiler, linker, debugger, assembler, and commercial Verilog simulator. Verilog is a hardware description language (HDL) used by engineers to describe hardware. The fabric simulator includes cycle-accurate models of all the hardware subsystems (fabric, memory controller, configuration controller, and so on) except for the embedded ARC processor. The ARC processor is modeled using an instruction set simulator (ISS). The GNU debugger serves as the front end for the simulator, and a graphical environment is available via the use of the data display debugger (DDD).

The designer starts by partitioning the system (software for the RCP) into functions that run on the ARC processor and functions that run on the fabric. The ARC is usually reserved for control and off-line processing, while the fabric is used for high-speed signal processing streaming type functions. The designer then models the fabric-partitioned algorithms using the C language; at the same time the C model is used to create a series of test benches. A test bench normally includes a set of input test vectors and the accompanying output test vectors. The designer then uses Chameleon's reconfigurable design entry language (similar to assembly language) to create kernels that are equivalent to the modeled C functional blocks. The C-Side assembler then automatically converts the kernel descriptions into standard Verilog. The generated Verilog code can then be verified by running it through a commercial Verilog simulator with the previously generated test-bench input vectors and comparing the output vectors with those produced by the C model. For the next phase of design the Verilog kernels are processed and synthesized by the C-Side V2B (short for Verilog to bits) tool. The result is a bit file that is used to configure the RCP. The designer then uses the compiler and linker to integrate the application-level C code with the Verilog kernels and the rest of the standard C function; this is a transparent process. Now the designer can use the full RCP simulator to verify that the overall C code is meeting the system-level specifications.

6.4.2 Adaptive Computing Machine

Another start-up company from San Jose, California, is QuickSilver Technology, created in 1998 by former Xilinx executives. This company has developed a reconfigurable processor designed to be embedded into wireless terminals and mobile devices. This new signal processing integrated circuit is called an Adaptive Computing Machine. It is intended to be as programmable as a DSP and consume as little power as an ASIC.

The technology is based on two techniques. The first component involves creating a custom data path that exactly fits the best sequence of instructions to implement the required algorithm. The design of the data path can be stored as software and loaded onto the device quicker than traditional downloads from firmware ROM; also, the data path can be reconfigured in the same way hundreds of thousands of times a second ($\approx$ microseconds) [15]. This reconfiguration speed is not as quick as an RCP ($\approx$ nanoseconds) but is much faster than an FPGA (fractions of a second). As an example, assume that a 27-input floating-point addition was performed by a DSP with a floating-point MAC unit; the DSP may consume 30–50 cycles to perform this type of operation. For the same function the ACM would download and configure in programmable hardware a 27-input adder just prior to being required by the data. Assuming that the data is ready in the appropriate part of the hardware, the ACM can complete the operation in only seven cycles.

Now consider the problem of implementing a 24-tap FIR filter; the ACM has the capability to program the hardware with a data path complete with 24 multipliers and 24 static coefficients. Multipliers are big consumers of silicon real estate and this is where the second component of the technology produces efficiency gains. In the ACM's programmable logic, a full integer multiplier can be replaced by a modest amount of random logic. Also, for additions that involve constants (FIR taps), simple combinatorial circuits can replace full adders. This concept is similar in nature to the implementation of CDMA correlations, as discussed in Section 6.2 (i.e., multiplication can be replaced by addition and subtraction). The result is significantly less hardware and therefore lower power consumption.

6.5 FPGAs

An FPGA is an integrated circuit containing a large array of identical logic cells; there can be as many as 15,000. The interconnects between cells are programmable; changes can be made in the field over the lifetime of the device. Cells can be grouped into logic blocks to form higher-level functions (e.g., lookup tables, multiplexers, flip-flops, multipliers, and RAM). Each logic block can implement a high degree of combinatorial logic functions on a given set of inputs. Functionality is usually described with a hardware language such as VHDL or Verilog; a synthesis tool is then used to automatically determine the routing required through the device and produces a register transfer logic (RTL) description. The combination of the algorithm being implemented, the VHDL code and the efficiency of the synthesis tool will determine the resources used and the maximum clock frequency for the FPGA. This is a major difference when compared with a DSP or RCP, where the maximum clock frequency is fixed and independent of the code running on the device. Implementation in FPGA can result in a clock frequency significantly less (by 50%) than the theoretical maximum and device utilization of approximately 50%.

Xilinx and Altera are the two biggest suppliers of FPGA products. Xlinix's range includes Spartan, Vertex, and Vertex II, with Vertex II providing the largest signal processing capacity. Altera's more powerful devices include the Mecury, Apex 20k, and Apex II products.

The number of logic gates on a single FPGA can be very large (e.g., the Xilinx Vertex II device ranges from 40,000 to 8 million system gates; see Table 6.6). These large devices can have 168 on-chip 18-bit by 18-bit multipliers and 1,100+ I/O pins with efficient code and tools; they can be clocked at high speed.

Even though FPGAs are programmable and potentially very powerful, they have not taken the place of DSPs. FPGAs are more difficult to program efficiently and their inability to dynamically reconfigure means that they are

Table 6.6
Typical Xilinx FPGA Parameter Ranges

Device	System Gates	Row by Column	Slices	Multipliers	Max RAM Kb	I/O Pins
XC2V40	40k	8 × 8	256	4	72	88
XC2V1000	1M	40 × 32	5,120	40	720	432
XC2V4000	4M	80 × 72	23,040	120	2,160	912
XC2V6000	6M	96 × 88	33,792	144	2,592	1,104
XC2V8000	8M	112 × 104	46,592	168	3,024	1,108

less than ideal targets for object-oriented software and the ideal software architectures described in Sections 7.3.1 and 7.3.2.

Traditionally, FPGAs have been used for prototyping designs ahead of implementation in fixed-function ASICs. The FPGA development tools that are available to map algorithms to silicon are efficient in engineering time but not necessarily in device resources. This is consistent with the theme of prototyping, where functionality and time are of the essence and the costs of the raw materials are usually considered negligible. In this case the thousands or perhaps tens of thousands of dollars of FPGAs will be outweighed by the hundreds of thousands of dollars and months of engineering time saved. To achieve better efficiency the designer must optimize by hand; this requires an intimate knowledge of the hardware and many years of experience.

Theoretically the largest FPGA can be clocked at 250 MHz using all its 168 multipliers; this yields an incredibly large MMACS figure of 42,000. This compares with 8,800 for a DSP and 2,400 for an RCP. However, if the actual clock frequency only reaches 50% of the maximum, and resource utilization also only reaches 50%, the resultant MMACS capacity quickly diminishes by 75% to 10,500 MMACS, a figure much closer to a DSP. Very large FPGA devices are difficult to manufacture and, consequently, extremely expensive when compared with DSPs. The largest FPGAs are priced similarly to DSPs when comparing dollars per theoretical MMACS; however, this situation rapidly deteriorates if actual signal processing capacity only reaches 25% of the maximum; for this case the FPGA becomes four times more expensive.

Despite the disadvantages with FPGAs they have still found their way into several software radio designs, an example of an FPGA implementing chip-rate processing for UMTS is provided in [16]. FPGAs are also expected to be popular choices for less cost-sensitive radios, that is, military SDR.

6.6 Symbol Rate and Chip-Rate Partitioning

Assuming that the designer is willing to support more than one programming language and tool set, a sensible partitioning choice could allocate chip-rate processing to a reconfigurable processor and symbol rate processing to a DSP. By using a high-bandwidth connection between the two, the partitioning can be refined and modified during the design phase with the potential for some chip-rate functions to be moved onto the DSP and parts of the symbol-rate processing to the reconfigurable processor. Chapter 8 touches on suitable interprocessor communications designs.

Although the DSP may not be capable of high-capacity chip-rate processing (on a single device), it is possible to implement restricted functionality; for guidance on implementing a UMTS Rake receiver, please refer to [17].

6.7 Conclusion

The realm of the DSP is now being challenged by many new and innovative signal processing devices. Massive capital funding is being injected into these technologies, and software radio will benefit as a result. Each device targets a different set of problems, and the newer processors specifically set out to overcome the weaknesses of the more traditional architectures. For some applications one device will stand out among the rest; for others the choice will not be as clear cut. Theoretical metrics are a good way to initiate an understanding of a device's capability; however, the final judgment is best made by benchmarking the actual algorithms and paying close attention to the efficiency of the compiler and its ability to optimize software.

Chapter 7 covers the software used to program the devices from this chapter, including a more detailed description of the hardware languages of VHDL and Verilog.

References

[1] Gilhousen, K. S., R. Padovani, and C. E. Wheatley, "Method and System for Providing a Soft Handoff in Communications in a CDMA Cellular Telephone System," U.S. Patent 5101501, March 31, 1992.

[2] Price, R., and P. E. Green, Jr., "A Communication Technique for Multipath Channels," *Proceedings of IRE*, Vol. 46, March 1958, pp. 555–700.

[3] Lee, J. S., and L. E. Miller, *CDMA Systems Engineering Handbook*, Norwood, MA: Artech House, 1998, pp. 972–982.

[4] Bruck, G. H., and P. Jung, "Software Defined Radio in Drahtlosen Endgeraten," University of Duisberg, March 29, 2001.

[5] Rabaey, J. M., "Beyond the Third Generation of Wireless Communications—ICICS99 Singapore," University of California Berkley, 1999.

[6] http://www.chameleonsystems.com.

[7] Texas Instruments, "TMS320C55x Technical Overview SPRU393," February 2000.

[8] Analog Devices, "ADSP-TS001 Preliminary Technical Data," REV PrC, December 1999.

[9] Analog Devices, "ADSP-TS101-S TigerSHARC DSP," 2001.

[10] Johnson, W., *Superscalar Processor Design*, Technical Report CSL-TR-89-383, Computer Systems Laboratory, Department of Electrical Engineering and Computer Studies, Stanford University, June 1989.

[11] Motorola, "Fact Sheet MSC8102," MSC8102FACT/D, 2001.

[12] http://www.ti.com.

[13] http://www.metrowerks.com.

[14] Chameleon Systems, "Wireless Base Station Design Using Reconfigurable Communications Processors," V1.0 0005, 2000.

[15] QuickSilver Technology, "Technology Backgrounder," Version 1.6, 2000.

[16] Altera, "Implementing a W-CDMA System with Altera Devices and IP Functions," AN129 Version 1.0, September 2000.

[17] *Implementation of a WCDMA Rake Receiver on a TMS320C62x DSP Device*, Texas Instruments Application Report SPRA680, July 2000.

7

Software Architecture and Components

In previous chapters we have focused on system-level design, hardware selection, and functional partitioning. These topics currently dominate the early design stages of a software defined radio project. This is particularly true for 3G cellular mobile radio, because state-of-the-art hardware is required to meet demanding performance specifications. Ultimately it is the software that provides the functionality, and the software architecture must include characteristics and mechanisms that allow for an efficient utilization of the underlying hardware platform.

7.1   Introduction

Significant effort is being invested in developing open software architectures and interfaces; this work aims to foster software reuse, portability, and compatibility. Many in the SDR industry are hoping that an open source paradigm will produce many of the benefits similarly experienced by users of the LINUX operating system. This chapter introduces two organizations that are helping to define the open software radio platform and follows up with some of the technologies, including operating systems and software languages, that must be chosen during the project's detailed design phase.

7.2 Major Software Architectural Choices

7.2.1 Hardware-Specific Software Architecture

In Chapter 1 we introduced the concept of the ideal software defined radio (see Figure 1.2) and proposed a layered abstracted software architecture. This concept allows the application software to be independent of an underlying standardized hardware platform. The aim of this approach is that ultimately any investment in the development of application software is maintained when ported to new (and presumably better) hardware platforms that comply with the standard.

Commercially designed cellular mobile radio equipment (terminals and base stations) has traditionally been developed as a black box. A software-level interface is not provided; only high-level functional and physical interfaces are exposed (e.g., mobile phone serial interface or BTS Abis interface [1]). In most cases radio equipment (1G and 2G) from different vendors will have incompatible software architectures, where the driving requirements have been to support legacy hardware or software or both.

The degree to which these traditional developments have utilized common interfaces and object-oriented (OO) design is difficult to gauge, because details of the developments are most often kept in-house.

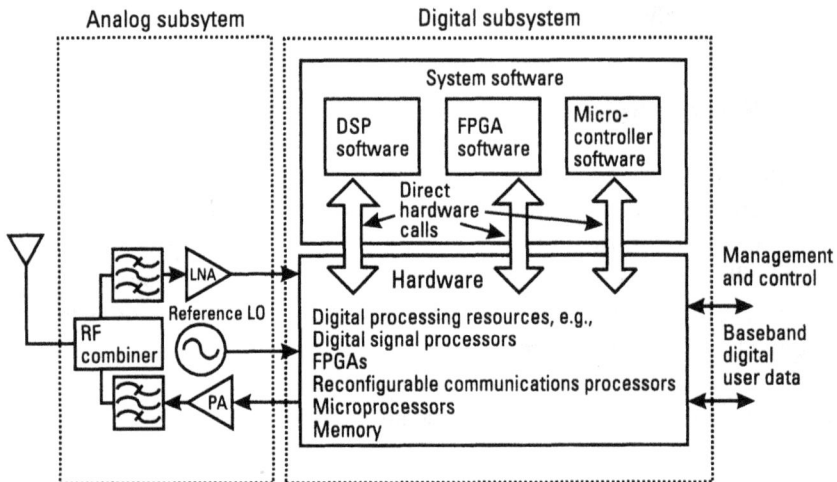

Figure 7.1 Hardware specific software defined radio architecture.

The system architect could choose a hardware-specific software architecture (see Figure 7.1) and still meet the requirement of implementing a software defined radio. For this case the system software is developed in a native language (to the processor), and the software makes direct calls to the hardware resources (e.g., direct manipulation of registers or I/O). This approach is used in conjunction with a structured design method, such as data flow diagrams, to capture the software design. The resultant architecture is not object oriented and certainly not portable.

Therefore, the ideal SDR, as detailed in Chapter 1, should include an emphasis on object orientation (i.e., the radio will be under the control of software and designed to use as much object orientation and software reuse as possible).

7.2.2 Abstracted Open Software Architecture

The term "bloatware" is part of the PC software lexicon as a result of inefficient object-oriented implementations. These poorly designed applications and operating systems are characterized by a tendency, following upgrade, to consume far more hardware resources (RAM, disk space, and CPU cycles) than the previous version for questionable improvements in functionality. As an example, the Microsoft NT operating system reportedly expanded from 16 million lines of code in version 4 to approximately 40 million lines of code in version 5.

The question then becomes: Can a real-time system such as a 3G software radio afford to become as fully object oriented as a PC application? Also, can the abstraction extend from the system's general-purpose microprocessors (e.g., Intel, Motorola 82xx), where the control and signaling software is often hosted to flow across the radios' embedded processors (e.g., DSP, FPGA, RCP, *u*P)? Lynes [2] recognizes this question by stating that traditionally there has been difficulty in providing a similar level of abstraction in the more dedicated resources used to process the physical layer signals." Figure 7.2 illustrates Lynes's [2] approach for a 3G SDR, where a "thin" layer of abstraction is spread across the layer one processing resources.

Figure 7.2 presents both a functional view of the radio's layer one processing and a software architectural view. The shaded areas [e.g., DSP (AP) and FPGA (AP)] make up the application software and the other blocks (e.g., PDC/PUC Lib) constitute the middleware. The level of abstraction provided by this middleware becomes thinner for the highest-rate signal

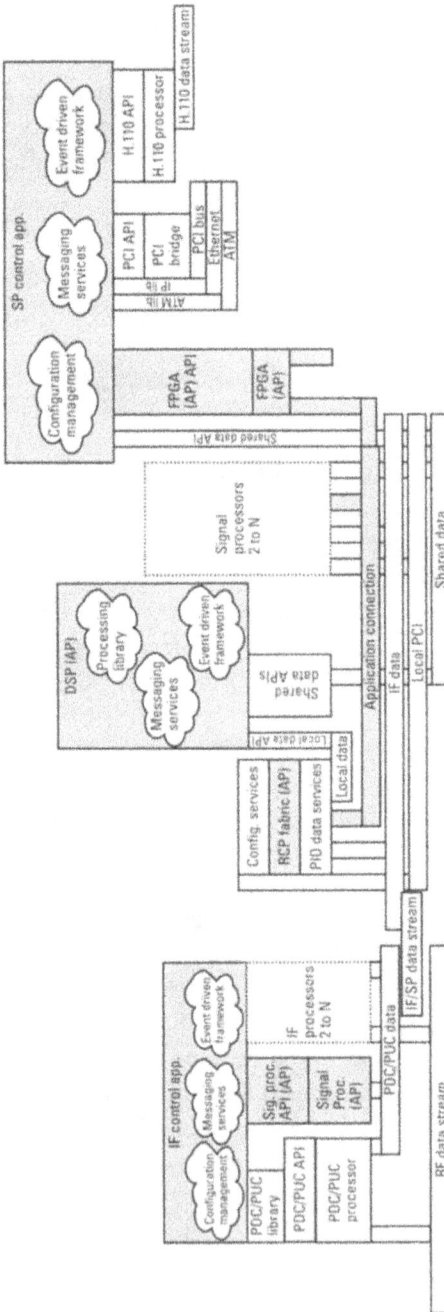

Figure 7.2 Layer one abstraction. (*Source*: ACT Australia, 2001. Reprinted with permission.)

processing elements, since efficiency and channel density are still major objectives. The IF processor library includes standard definitions for all common air interface standards; however, the control application, which is responsible for monitoring gain control, timing offsets, and synchronizing parameter changes, is application dependent and must be developed. The framework (middleware and hardware) provides an event-driven control structure, where the triggering events, much like a sensitivity list, can be modified to suit the application. APIs are provided to access all measurable and controllable parameters.

A similar control framework is provided for the baseband processor. All local and board to board interprocessor communications are handled by middleware. Application code is also needed for the processing element RCP and DSP devices. Both these devices are supported by application libraries and code framework that take care of all the data transfer and code loading functions. The application exists in a generic buffer-in, process, buffer-out environment.

7.3 Software Standards for Software Radio

There are two key organizations pushing forward the adoption of software standards for the development of software defined radios.

The Software Communications Architecture Specification (SCAS) [3] is published by the United States Joint Tactical Radio System (JTRS) Joint Program Office (JPO). This defense program has set goals for future communications systems (i.e., to increase flexibility and interoperability; ease up grade ability; and reduce acquisition, operation, and support costs). The JTRS states that the SCAS is not a system specification but a set of rules that constrain the design of systems to achieve these objectives. The U.S. government expects the basic SCAS to become an approved commercial standard through the Object Management Group (OMG) and has designed the specification to meet commercial as well as military application requirements. The OMG is becoming more involved in software radio specification and has created a special interest group, see Section 7.3.3 for more information.

The other effort is led by the SDR Forum's Mobile Working Group, which has developed a Distributed Object Computing Software Radio Architecture (DOCSRA) [4]. The development describes a software framework for open, distributed, object-oriented, software-programmable radio architectures. The Mobile Working Group developed the DOCSRA using the IEEE's recommended methodology [5] for architectural descriptions.

The SCAS and DOCSRA have a similar heritage with some common terms (e.g., *FileManager* and *DomainManager*); we cover both proposals due to their importance and potential to become industry standards. The presented architectures are object oriented and the designs are captured using Unified Modeling Language (UML) [6] notation. The standardization and open architecture approach is already attracting attention, with early signs of references in commercial product [7].

7.3.1 JTRS Software Communications Architecture Specification

The software framework of the SCAS defines the operating environment (OE) and specifies the services and interfaces that applications use from that environment. The OE is comprised of the following entities:

- A core framework (CF);
- CORBA middleware;
- A POSIX-based operating system (OS) with associated board support packages.

7.3.1.1 Architecture Overview

A graphical depiction of the relationships between the OE (CF and COTS) and SDR noncore components is provided in Figure 7.3; this details the key elements and IDL interfaces for the CF.

The SDR software structure shows that the noncore components are abstracted away from the underlying hardware, and all entities are connected via a logical software bus by using CORBA. Adapters are provided to allow non-CORBA modem, security, and I/O components to interface with the CORBA components via the CORBA bus.

The software architecture is capable of operating using COTS hardware bus architectures (e.g., VME, cPCI, and so on); however, the actual implementation will be determined by the derived performance requirements of the noncore components (e.g., data bandwidth and timing).

COTS operating systems (OSs) with real-time embedded capabilities such as preemptive multitasking are expected to be suitable for SDR. The OS is also assumed to be portable operating system interface (POSIX) compliant and the SCAS recommends the use of POSIX 1003.13 [8]. POSIX was first published by the IEEE in 1990 and now consists of more than 30 standards [9], ranging from basic OS definitions through to those with advanced real-time extensions. These extensions include such features as

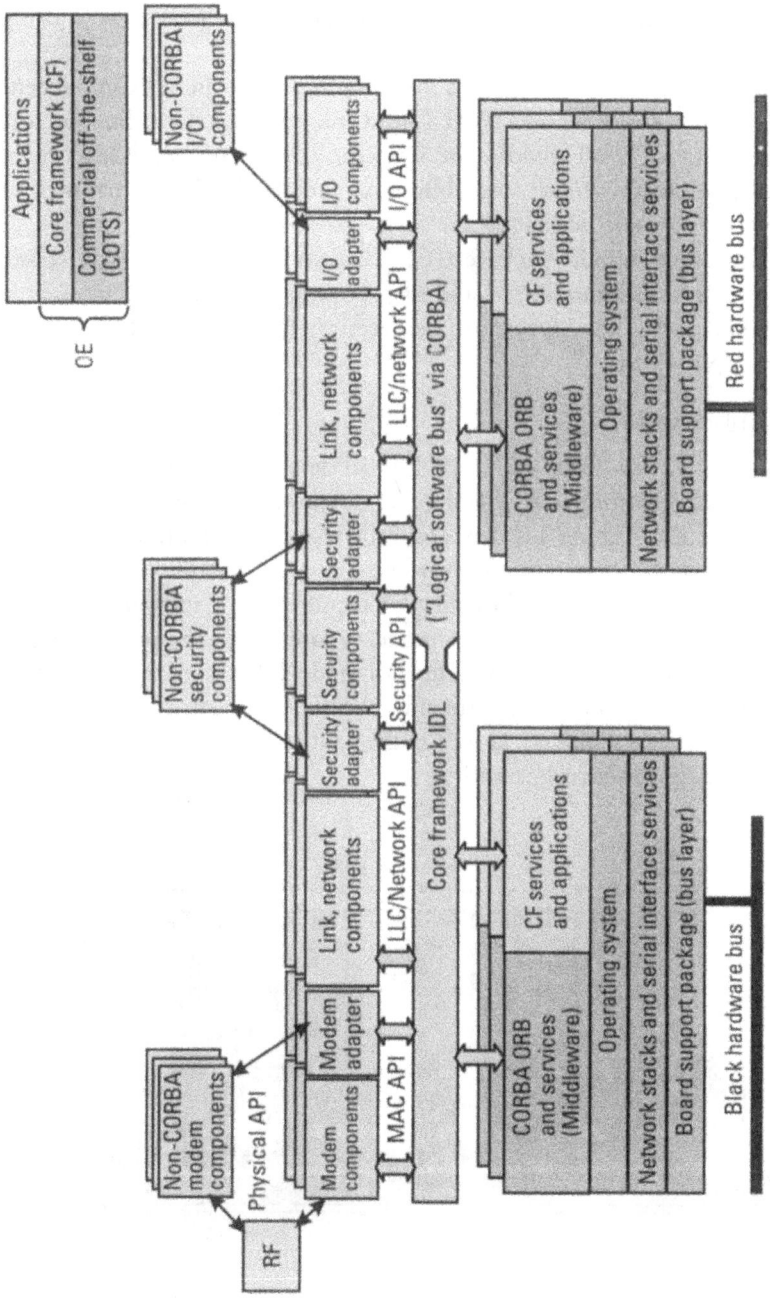

Figure 7.3 Software structure. (*Source: JTRS JPO, 2001. Reprinted with permission.*)

threads, priority scheduling, and semaphores. POSIX 1003.13 does not contain any additional features; instead, it groups the functions of existing POSIX standards into units of functionality.

The operating environment is the integration into an SDR implementation of the CF services and COTS infrastructure (e.g., OSs, bus support packages, and CORBA middleware services). The SCAS emphasises open standards, multiple vendor available commercial (COTS) elements, and higher-order software languages.

The board support package (BSP) dual-connection to a black secure bus and a red nonsecure bus would only be implemented for defense type applications and is not expected to be necessary for civilian and 3G mobile cellular applications.

7.3.1.2   Functional View

The functional view starts with a traditional description of the system with data flow and control paths. The data flow follows the convention adopted in this book (e.g., Figures 1.2, 2.3, and 7.1), where the air interface is on the left and the network interface on the right.

This traditional data flow view is captured as a software reference model, depicted in Figure 7.4. The model is based upon the programmable modular communications system reference model [10]. It serves to introduce the various functional roles performed by the SDR software entities (without dictating a structural model), as well as the control and traffic data interfaces between functional software entities.

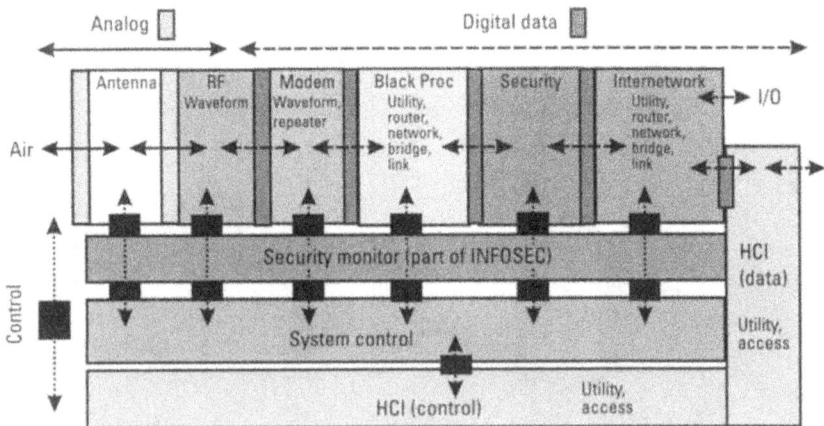

Figure 7.4   Software reference model. (*Source:* JTRS JPO, 2001. Reprinted with permission.)

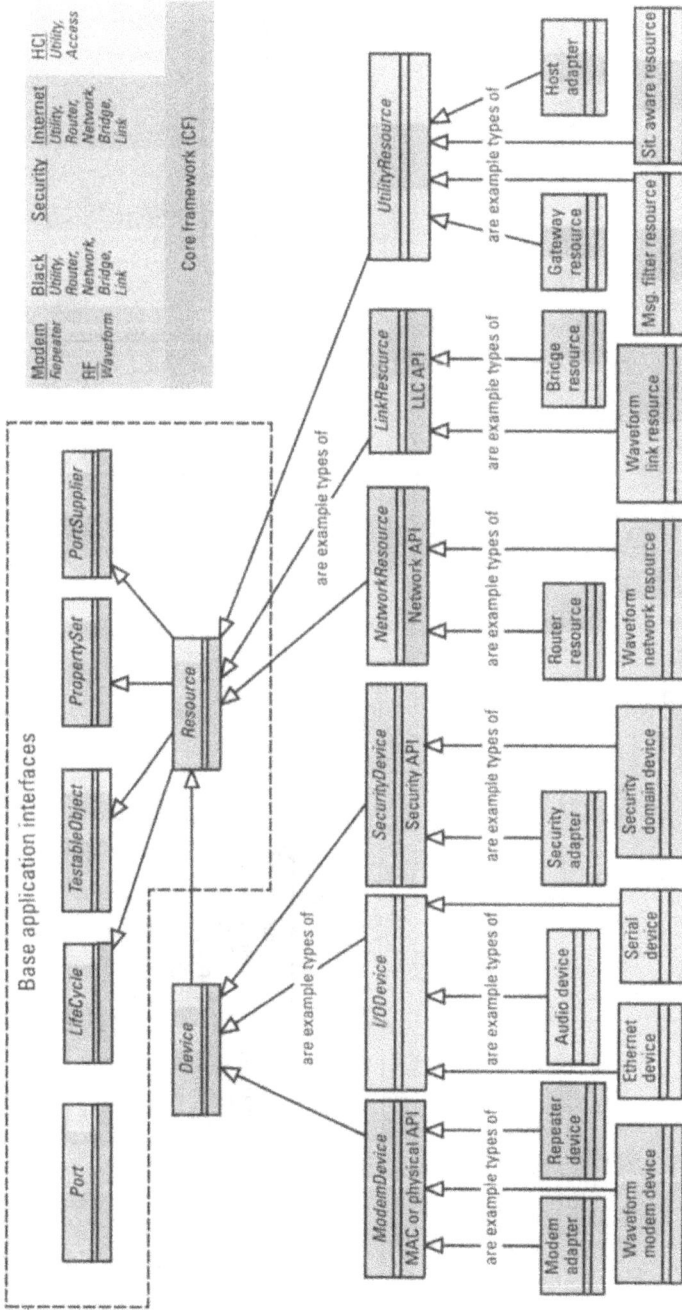

Figure 7.5   Conceptual model of Resources. (*Source:* JTRS JPO, 2001. Reprinted with permission.)

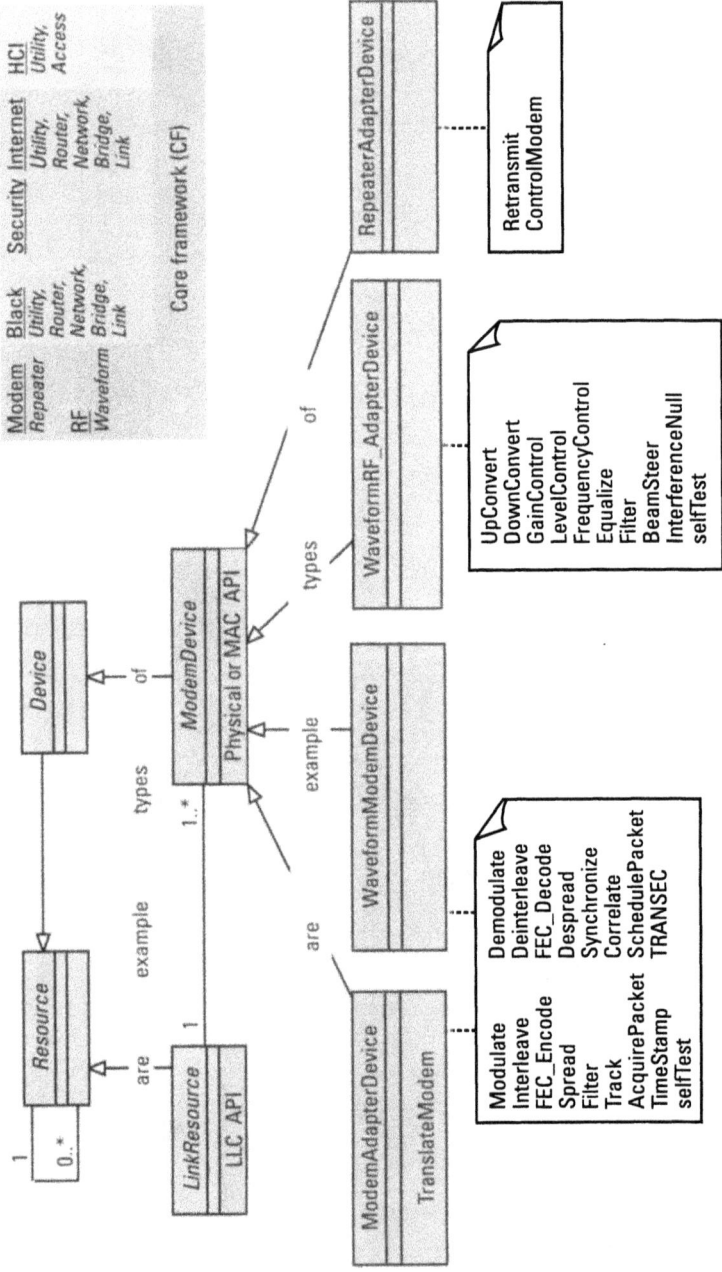

Figure 7.6 ModemDevice conceptual class diagram. (*Source:* JTRS JPO, 2001. Reprinted with permission.)

The remainder of the SCA specification extends this functional view by capturing the software architecture in object-oriented models. The SCAS realizes the software reference model by defining a standard unit of functionality called a *Resource*. All applications are comprised of *Resources* and *Devices* (e.g., the *ModemDevice* includes the antenna, RF, and modem entities) that are types of Resources used as software proxies for actual hardware devices. Figure 7.5 shows examples of implementation classes for *Resources*.

The parent *Resource* is extended by its children in the following ways: *ModemDevice* adds physical devices common to all modems, *LinkDevice* adds link layer interfaces, *NetworkResource* adds network layer interfaces, *I/ODevice* adds a set of input/output devices such as Ethernet and serial, *SecurityDevice* adds security devices, and the *UtilityResource* adds utilities such as message filtering and a network gateway interface. The base application interfaces encapsulated by the *Resource* class provides a mechanism for pushing or pulling messages (*Port*) between *Resources* and *Devices*.

A conceptual class diagram for a *ModemDevice* is provided in Figure 7.6. The *ModemDevice* provides a standard for the control and interface of a modem and encapsulates diverse implementations of smart antenna, RF, and modem functions. Figure 7.6 also illustrates that a parent resource can be extended by adding a child (e.g., *WaveformModemDevice*) with more functionality.

7.3.1.3 Networking Overview

The SCAS includes specification of the external protocols that define communication between an SCAS-compliant software radio and its peer systems. Although the SCAS references many military protocols, it also considers the popular IS-95A CDMA mobile cellular standard as an example.

Figure 7.7 illustrates how the SCAS APIs map onto the OSI seven layer networking model. This mapping is not dissimilar to that used in the cellular mobile world, where the UMTS (3GPP) and CDMA2000 (3GPP2) specifications concentrate on the physical (layer one), link (layer two), and networking (layer three) layers.

7.3.1.4 Core Framework

The core framework (Figure 7.8) is part of the OE and is the essential core set of open application layer interfaces and services to provide an abstraction of the underlying software and hardware layers for software application designers. The CF consists of the following:

- Base application interfaces (*Port, LifeCycle, TestableObject, Property-Set, PortSupplier, ResourceFactory,* and *Resource*), which can be used by all software applications

- Framework control interfaces (*DomainManager, DeviceManager, Application, ApplicationFactory, Device, LoadableDevice, ExecutableDevice,* and *Aggregate*), which provide control of the SDR

- Framework services interfaces, which support both core and non-core applications (*FileSystem, File, FileManager,* and *Timer*)

- A domain profile, which describes the properties of hardware devices (device profile) and software components (software profile) in the SDR

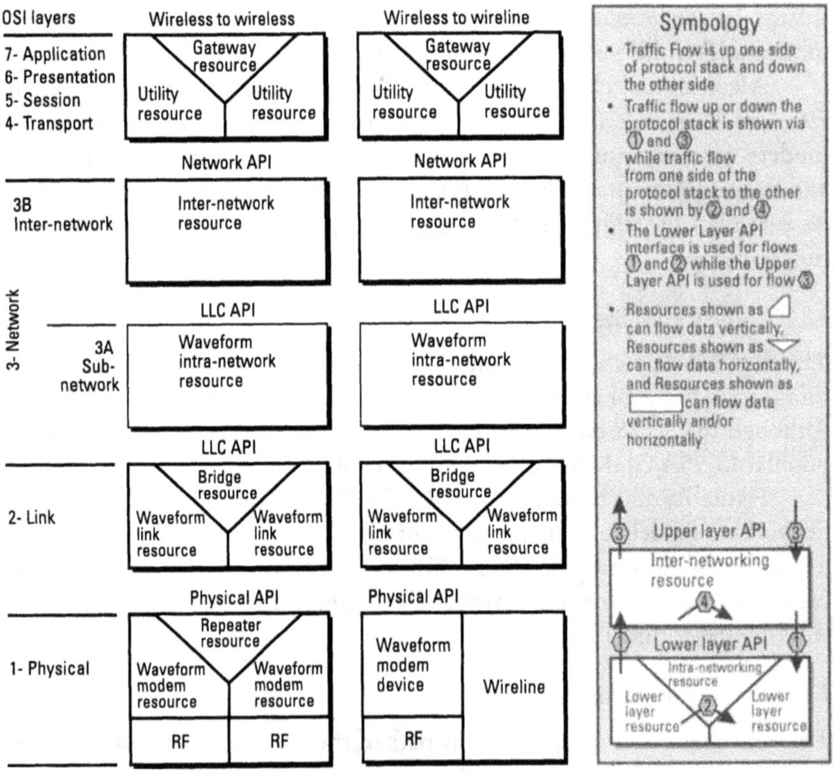

Figure 7.7   Networking mapped to OSI network model. (*Source:* JTRS JPO, 2001. Reprinted with permission.)

As in a UML class diagram, the rectangles in Figure 7.8 represent classes, and the lines connecting classes represent associations. Associations are assumed to be one to one unless annotated as one to many (1 . . . 1 . . .*) or many to many (1 . . . * 1 . . . *); in our example, the *DomainManager* oversees one or many *DeviceManagers,* where the many set includes the number 1. Association lines can have an end adornment; a hollow diamond indicates an aggregation type of relationship (e.g., *FileSystem* is part of *DeviceManager,* and *DeviceManager* is part of *DomainManager*). A triangle at the end of an association line represents a generalization or parent/child relationship. As an example, an *Application* is a type of *Resource,* and as such the *Application* inherits its behavior and interfaces from its parent *Resource.*

The SCAS provides a detailed description of the core framework interfaces and operations. The OMG-developed Interface Definition Language (IDL) is standardized by the International Organization for Standardization as ISO/IEC 14750 [11] and is used to describe these interfaces. According to the CORBA specification [12] an interface is a description of the set of possible operations a client may request of an object. As per the OMG's objectives, the IDL language is independent of compiler, software language, or operating system.

The *DomainManager* (as depicted in Figure 7.9) is a key CF application; its interfaces can be logically grouped into the categories of human computer interface (HCI), registration, and CF administration. The HCI operations are used to configure the domain, get the domain's capabilities (devices, services, and applications), and initiate maintenance functions. Host operations are performed by an HCI client capable of interfacing to the *DomainManager.* The registration operations are used to register/unregister *DeviceManagers, DeviceManager's Devices, DeviceManager's Services,* and *Applications* at startup or during run time for dynamic device, service, and application extraction and insertion. The administration operations are used to access the interfaces of registered *DeviceManagers* and *DomainManager's FileManager.*

The sequence of operations involved in a *DeviceManager* registering with the *DomainManager* is depicted in Figure 7.10. The sequence starts with the registerDeviceManager operation from the *DeviceManager* to the *DomainManager* and completes with the writeRecord operation from the *DomainManager* to the *Log.*

The other major CF interface is the *DeviceManager;* its job is to manage a set of logical devices and services. Upon startup the *DeviceManager* registers itself with the *DomainManager* and uses a profile (deviceConfigurationProfile)

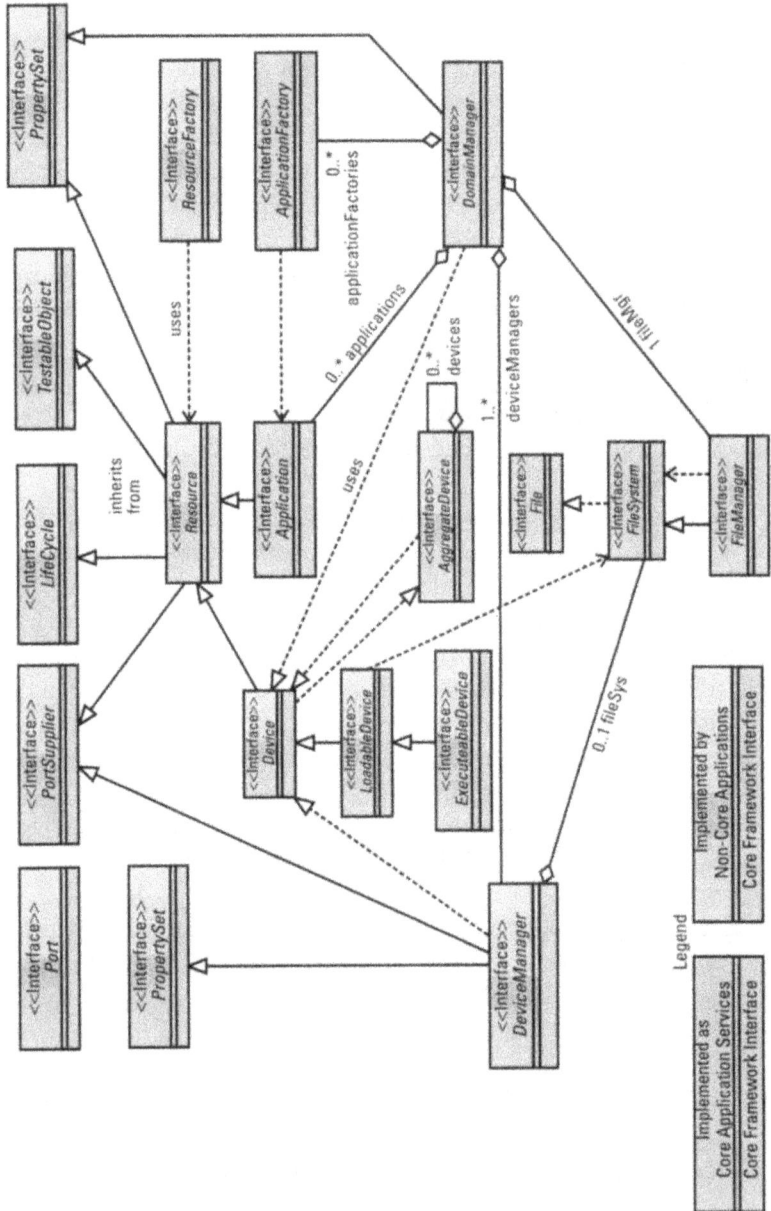

Figure 7.8   Core framework key elements. (*Source:* JTRS JPO, 2001. Reprinted with permission.)

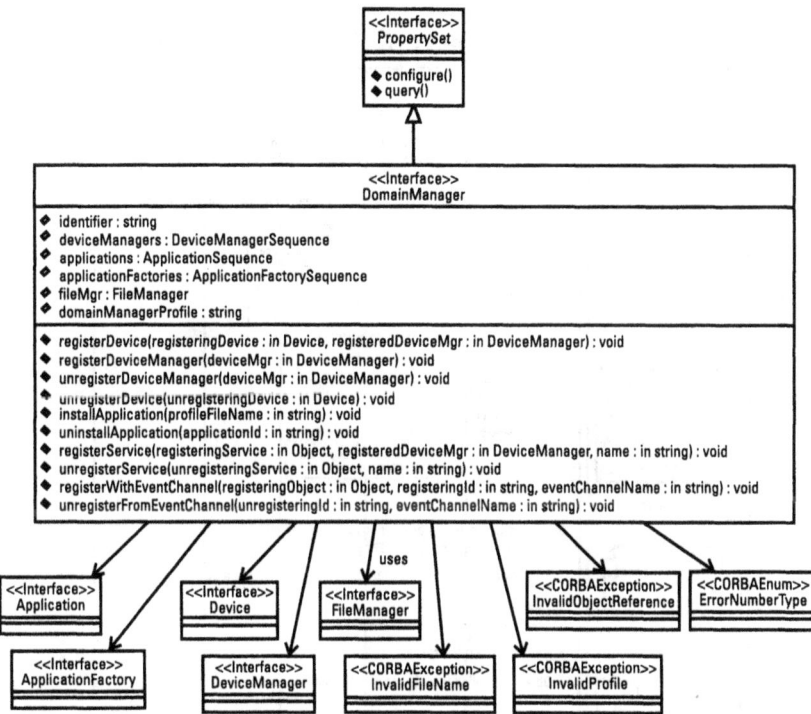

Figure 7.9 DomainManager interface UML. (*Source:* JTRS JPO, 2001. Reprinted with permission.)

attribute for determining the services to be deployed (e.g., *log*[s]), the *Devices* to be created and deployed, and the mount point names for *FileSystems*. It creates *FileSystem* components implementing the file system interface for each OS file in the system. The *DeviceManager* also initializes, configures, and starts logical *Devices* registered to itself. (See Figure 7.11.)

7.3.1.5 Hardware Architecture Definition

The SCAS provides guidance on partitioning the SDR hardware using an object-oriented (OO) approach. The OO method describes a hierarchy of hardware class and subclass objects that represent the architecture. Class structure is a hierarchy that depicts how object-oriented classes and subclasses are related. The class structure in the SCAS identifies functional elements that are used in the creation of physical system elements or hardware devices. As per the OO approach, devices inherit from their parents and share common physical and interface attributes; theoretically, this should make it easier to identify and compare device interchangeability.

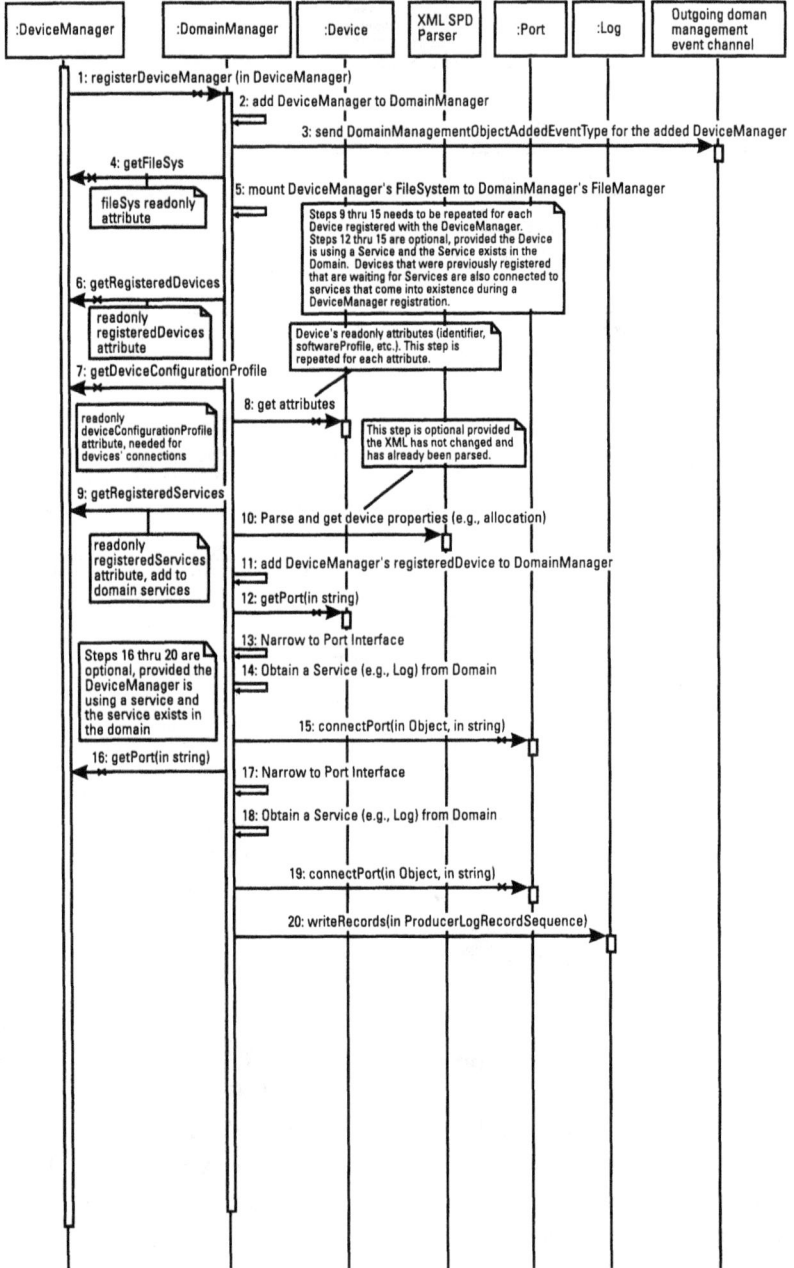

Figure 7.10 Example DomainManager sequence diagram. (*Source:* JTRS JPO, 2001. Reprinted with permission.)

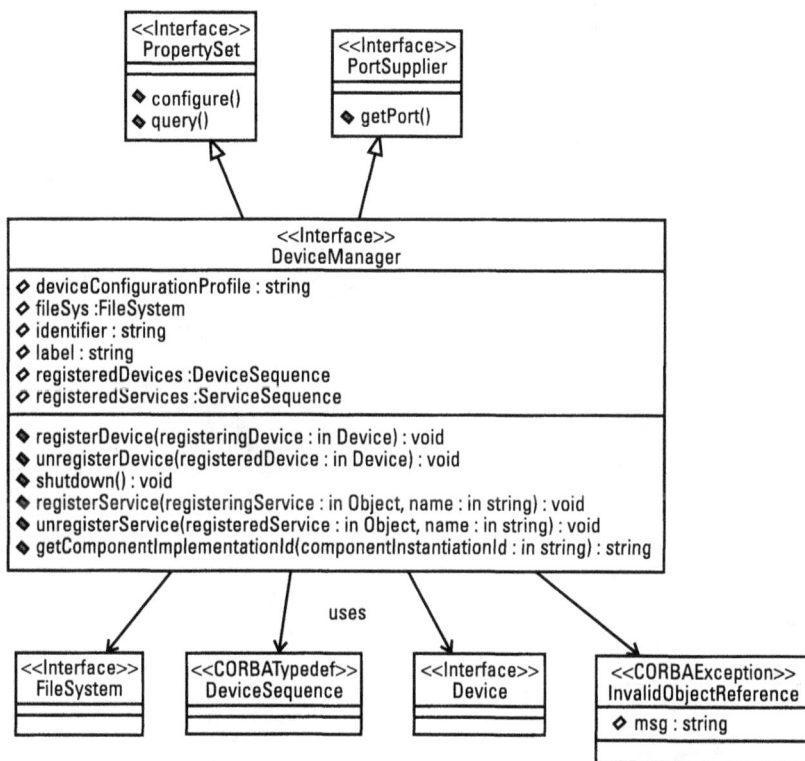

Figure 7.11 DeviceManager interface UML. (*Source:* JTRS JPO, 2001. Reprinted with permission.)

The overall hardware parent is the *SCA-Compliant Hardware* class; it defines attributes such as maintainability, availability, physical, environmental, and device registration parameters. *SCA-Compliant Hardware* has two child classes: *Chassis* and *HW Modules*. The *Chassis* subclass includes the attributes of module slots, form factor, back plane type, platform environmental, power, and cooling requirements. *HW Modules* is the parent to all module subclasses (e.g., *RF, Power Supply, Modem, GPS, Processor, Reference Standard*, and *I/O*).

Each of the hardware child classes can be further extended, and examples of the granularity of the extensions are depicted in Figure 7.12 for the *RF* class and Figure 7.13 for the *Modem* class.

The *RF* class is extended by the addition of *Antenna, Receiver*, and *Power Amplifier* child classes. The *Receiver* class includes many of the parameters discussed in previous chapters. The *Power Amplifier* child class

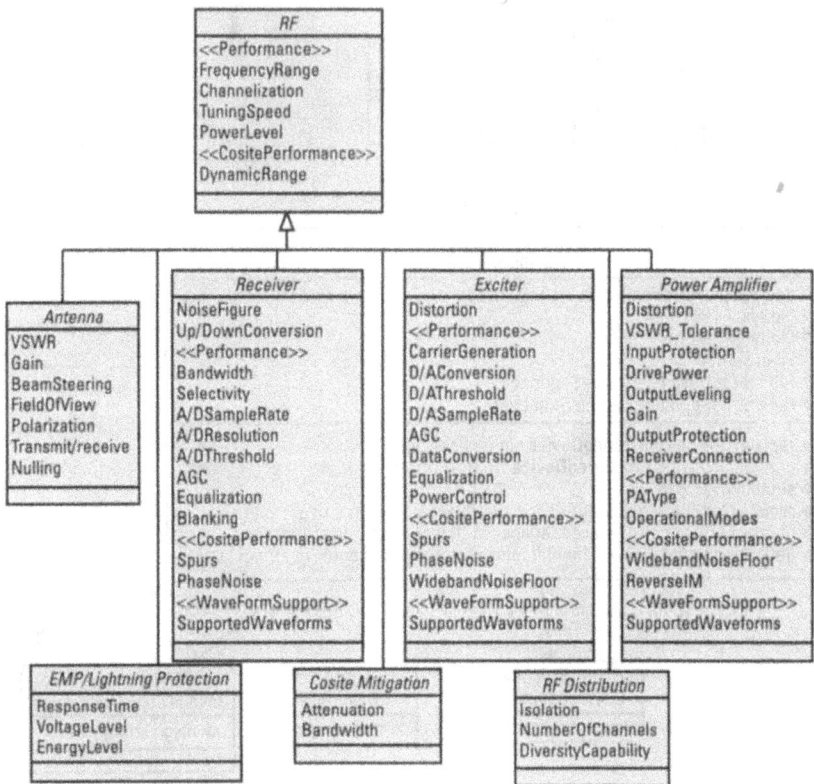

RF
<<Performance>>
FrequencyRange
Channelization
TuningSpeed
PowerLevel
<<CositePerformance>>
DynamicRange

Antenna
VSWR
Gain
BeamSteering
FieldOfView
Polarization
Transmit/receive
Nulling

Receiver
NoiseFigure
Up/DownConversion
<<Performance>>
Bandwidth
Selectivity
A/DSampleRate
A/DResolution
A/DThreshold
AGC
Equalization
Blanking
<<CositePerformance>>
Spurs
PhaseNoise
<<WaveFormSupport>>
SupportedWaveforms

Exciter
Distortion
<<Performance>>
CarrierGeneration
D/AConversion
D/AThreshold
D/ASampleRate
AGC
DataConversion
Equalization
PowerControl
<<CositePerformance>>
Spurs
PhaseNoise
WidebandNoiseFloor
<<WaveFormSupport>>
SupportedWaveforms

Power Amplifier
Distortion
VSWR_Tolerance
InputProtection
DrivePower
OutputLeveling
Gain
OutputProtection
ReceiverConnection
<<Performance>>
PAType
OperationalModes
<<CositePerformance>>
WidebandNoiseFloor
ReverseIM
<<WaveFormSupport>>
SupportedWaveforms

EMP/Lightning Protection
ResponseTime
VoltageLevel
EnergyLevel

Cosite Mitigation
Attenuation
Bandwidth

RF Distribution
Isolation
NumberOfChannels
DiversityCapability

Figure 7.12 RF class extension. (*Source:* JTRS JPO, 2001. Reprinted with permission.)

would need to be extended further for a wideband multicarrier radio and
may include additional parameters, such as peak to average power ratio (see
Section 3.6.1) or number of carriers. *Cosite Mitigation* implies careful
monitoring and control of interference-related parameters; this issue is
expected to be a significant one, especially during the deployment of cos-
ited 2G TDMA and 3G TDMA/CDMA systems. The SCAS notes that
antennas have been historically passive elements attached to the structure
that houses the communications system. In anticipation of technological
advances and "smart" antenna (see Chapter 9) functionality, *Antenna* is
included as an *RF* subclass with smart antenna parameters, such as Beam-
Steering and Nulling.

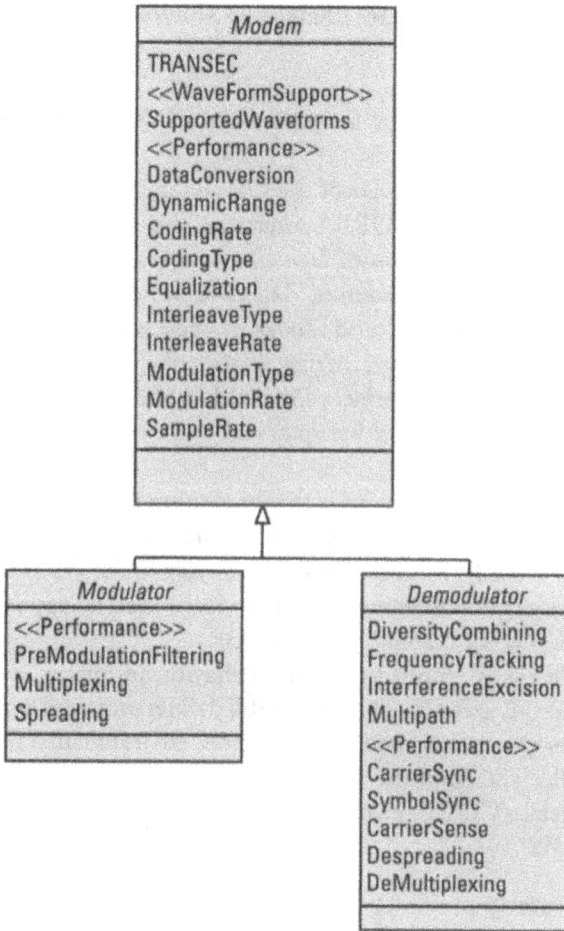

Figure 7.13 Modem class extension. (*Source:* JTRS JPO, 2001. Reprinted with permission.)

The modem class is extended into the transmit and receive subclasses of *Modulator* and *Demodulator*, as illustrated in Figure 7.13. For a multiple air interface mobile cellular radio the SupportedWaveforms attribute may have the valid values of GSM, IS 136, IS-95B, CDMA2000-1xRTT, or UMTS-FDD.

7.3.2   SDRF Distributed Object Computing Software Radio Architecture

7.3.2.1   Architecture Definitions

The architecture is defined by a core framework (CF), operating environment (OE), and rule set.

The core framework is part of the OE and includes a set of core applications, core services, base CORBA interfaces, and an optional core *Factory* interface. The core applications of *DomainManager* and *ResourceManager* are responsible for controlling resources. The core services of *FileManager, FileSystem, File, Logger, Installer*, and *Timer* provide support to the core and noncore applications. *Message, MessageRegistration, StateManagement*, and *Resource* are the CORBA interfaces inherited by the core and noncore software applications. Application life span can be controlled by including the core *Factory* interface.

The operating environment is the integration into an SDR implementation of the CF services and COTS infrastructure (e.g., OSs, bus support packages, and CORBA middleware services). Key to the OE is the inclusion of the software development environment (SDE) used by application developers to develop new capabilities for the SDR.

The rule set is embedded in the behavior, interfaces, and structure defined by the OE and CF. The DOCSRA defines an initial set of rules for the form factor (hardware standard), interfaces, environmental requirements, OSs, and SDE. The emphasis is on open standards, multiple vendor available commercial (COTS) elements, and higher-order software languages within the SDE.

7.3.2.2   Functional View

The functional view starts with a traditional description of the system with data flow and control paths. This traditional data flow view is captured as a software reference model and presented as a means of transitioning to an object-oriented model; it serves to define the functional view and broadly introduce the various functional roles performed by the SDR software entities. The object-oriented model is captured using the structural, logical, and use-case software views, as presented in the following sections.

The software reference model view is not meant to dictate a structural model of the elements. The DOCSRA advises that a software architecture that attempts to force-fit these functions into one place will not provide the flexibility demanded by the various SDR domains, or the reusability desired by software radio vendors and operators [13]. Previous chapters have mostly left the software architectural issues of an SDR alone. Instead, they have concentrated on defining air interface–specific functionality (antenna, RF, and modem) and performing realistic and efficient partitioning of that functionality considering 3G mobile cellular applications. The DOCSRA advice implies that the software architecture should be flexible enough to allow dynamic functional partitioning and use generic interfaces so that software can be ported from one hardware platform to the next.

An SDR will have several noncore user-oriented (or air interface–specific) software applications. These noncore applications can be considered as resources that inherit common types of behavior and common types of interfaces.

The *Resource* class is extended by its children: *ModemResource* adds physical devices common to all modems, *LinkResource* adds link layer interfaces, *NetworkResource* adds network layer interfaces, *AccessResource* adds a set of multimedia resources, *SecurityResource* adds services such as encryption and authentication, and the *UtilityResource* adds utilities such as message translation and network gateway interface.

The base class interfaces encapsulated by the *Resource* class provide a mechanism for the establishment (*MessageRegistration*) of message paths between resources, the actual path/pipe (*Message*) for communication, and a standard method for managing the resources states (*StateManagement*). Base class interfaces for a given resource (e.g., *ModemResource*) can be overlayed onto an embedded networking architecture by using a networking application program interface (NAPI).

The DOCSRA recognizes that a practical SDR will employ a variety of digital signal processing solutions using hardware (e.g., analog up-/downconverter), programmable hardware (e.g., digital up-/downconverter), and software (e.g., DSP, RCP, uP, and so on). Not all of these solutions will be CORBA-capable. To solve this problem the architecture includes a mechanism that allows a CORBA capable resource (e.g., *LinkResource*) to communicate with the noncore applications of the *ModemResource* (e.g., *WaveformModemResource*) via a standard modem NAPI interface. The mechanism that makes this possible is an agent (*ModemAgentResource*), which provides a transparent gateway between CORBA-capable and non-CORBA-capable resources.

7.3.2.3   Structural View

The structural view details the relationships between the OE (CF and COTS) and SDR noncore applications.

The SDR software structure ensures that the noncore applications are abstracted away from the underlying hardware and all entities are connected via a logical software bus by using CORBA. The recommended rule set emphasizes COTS hardware bus architectures (e.g., VME, cPCI, and so on). Real-time COTS operating systems (OSs) are proposed; these OSs are assumed to be POSIX compliant, with the DOCSRA recommending POSIX 1003.13 [8].

7.3.2.4 Logical View

The logical view of the DOCSRA provides a detailed description of the core framework interfaces and operations. As with the SCAS, IDL is used to describe the core framework interfaces.

The *DomainManager* is a key CF application; it has the job of object managing the software *Resources* and hardware components within the SDR. It is possible for software *Resources* (e.g., *ModemResource*) to directly control hardware components (e.g., DSP, RCP, or FPGA). The *DomainManager* allocates *Resources* to one or more *ResourceManager* objects; this decision is based upon many factors, including the visibility of the hardware to the *ResourceManager* and its availability. A prime goal of the DOCSRA is the use of generic resources and portability. This implies the ability to move hardware and software from one system to another without a change in the core framework; in a multiple air interface 3G BTS this may be interpreted as from one air interface to the next. The *DomainManager* is responsible for allocating hardware and software resources based on the functional requirements of the application or applications (in the case of a multiple air interface 3G radio). The *DomainManager* performs the allocation job by using a *DomainProfile*.

There is at least one *DomainManager* for each system, and the *DomainProfile* is used to store the information about the resources in the system. Physical resources such as digital frequency up-/downconverters and DSPs interact with the *ResourceManager* to report their availability; this information is logged into the *DomainProfile*.

The other CF application is the *ResourceManager*; its job is to boot, initialize, and report the capabilities of the hardware modules. The *deviceProperties*, *deviceList*, and *deviceExists* methods are available to the *ResourceManager* for communication back to the *DomainManager*. To ensure interoperability between modules within the SDR, a core list of properties should be defined; these can then be extended by module developers to increase usability as required. A key function of the ResourceManager is the capability to load and boot software on the various *Resources* it is responsible for managing.

7.3.2.5 Use-Case View

The final view of the architecture uses the UML use-case paradigm. The technique is designed to capture the totality of a system's actual interfaces and interactions. The design method was developed to reduce lost requirements and misinterpretations that result from informal design processes, where requirements may be discussed but not formally documented. The use-case diagram in Figure 7.14 is adapted from Figure 2.2-30 of the DOC-SRA [4]; it is a specific example for a potential 3G SDR BTS and the software development environment used to build it.

Within UML, a scenario [14] refers to a single path through a use case—one that shows a particular combination of conditions within that use case. Using the example of Figure 7.14, several scenarios would be developed for the boot up and initialize use case—one where boot-up runs smoothly, another where boot-up encounters problems, and so on. The use-case view of the DOCSRA follows this scenario convention (e.g., a "receive communications" and "transmit communications" scenario is provided for the equivalent of the "TX and RX control messages and traffic" use case).

7.3.3 The OMG

The nearly 800 member companies of the Object Management Group produce and maintain a suite of specifications that support distributed, heterogeneous software development projects from analysis and design through coding, deployment, run time, and maintenance. The specifications are written and adopted by using a well-defined open process. Any company, institution, or government agency can join the OMG and contribute to or influence the specifications.

The specifications are freely available for download from the OMG Web site (http://www.omg.org), and organizations are free to write software implementations that conform to the specifications and use them, give them away, or sell them. Such activities can be undertaken without OMG membership or license.

The OMG does not produce any software and concentrates purely on specification development. Software products implementing OMG specifications (e.g., UML, CORBA, and IDL) are available from hundreds of sources, including vendor companies and sources of freeware and open-source software.

The OMG also operates a number of special interest groups and has one group for software defined radio, please see http://swradio-omg.org/.

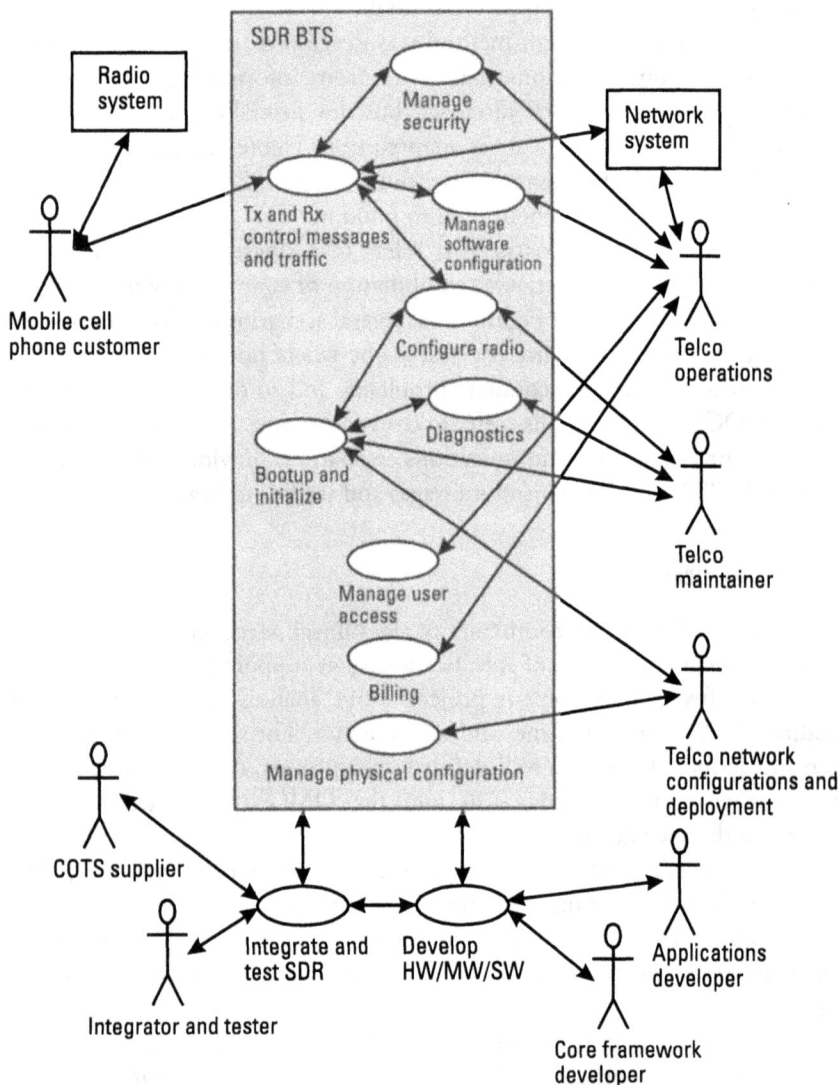

Figure 7.14 3G BTS SDR use cases.

The mission of this group is to:

- Communicate the requirements of the software radio community to the OMG and its various subgroups.

- Promote the use of OMG technologies within the software radio community by direct use of existing and emerging OMG specifications.

- Propose software radio-related extensions to existing and emerging OMG specifications by applying UML modeling techniques.

- Identify and propose new OMG specifications for the software radio community.

- Promote portability, reusability, scalability, and interoperability of software radio-based platforms and applications.

- Form liaisons with related organizations that share common goals and interests.

7.3.3.1 OMG CORBA

The Common Object Request Broker Architecture (CORBA) [12] is defined by the OMG as an open, vendor-independent architecture and infrastructure that computer applications use to work together over networks [15]. By using standard protocols the approach allows CORBA-compliant programs from one vendor to interoperate with those of another vendor, even if the underlying platform (processor, OS, language, and so on) is different. The OMG is responsible for developing and issuing what is now a suite of many documents. The specifications initially concentrated on connecting standalone computers across diverse networks; however, the latest CORBA includes an optional set of extensions called Real-Time CORBA (see Section 24 of [12]).

Each object in the system (e.g., Rake receiver) has its interface defined in Interface Definition Language (IDL). When a client invokes an operation (request) on the object, it must use the IDL interface to specify the operation and to marshal (serialize) the arguments that it sends. The same IDL interface definition is used once the invocation reaches the target object and the arguments are unmarshaled to allow the requested operation to be performed. Figure 7.15 illustrates the communication between software objects using object request brokers (ORBs). The client (e.g., a controller) is invoking different operations (e.g., addUser and setPilot) on two objects (*Rake*

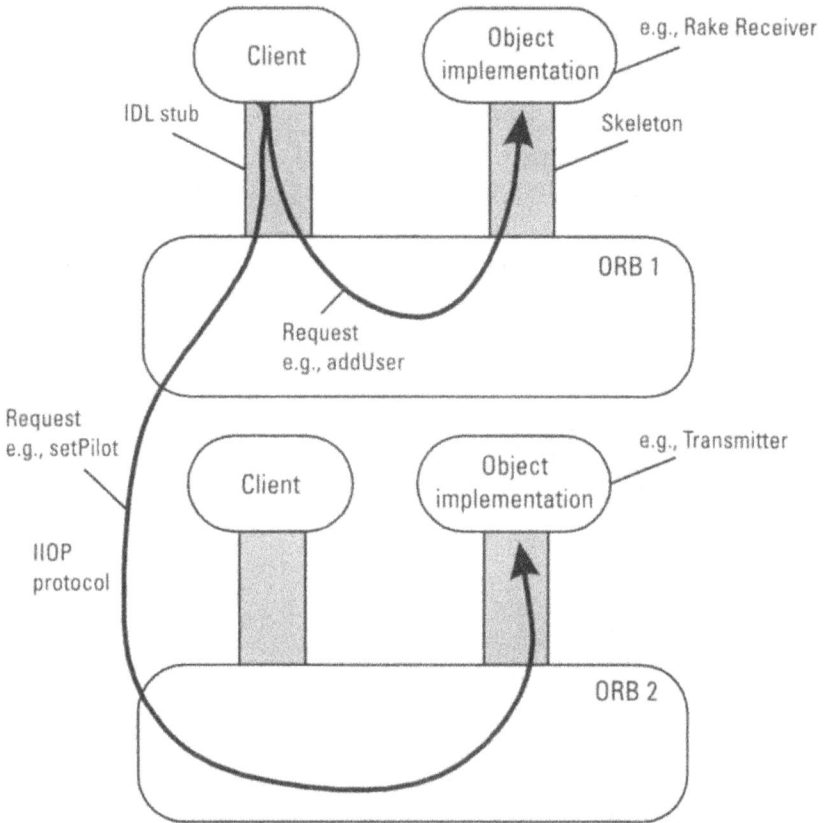

Figure 7.15 Communications via ORBs.

receiver and *Transmitter*) that are connected to different ORBs (ORB1 and ORB2).

Because CORBA strictly defines interfaces as well as the object implementation (code) and data being hidden (i.e., encapsulated), clients do not need to know where objects are or what the underlying platform for each is. The stubs and skeletons act as proxies for the clients and objects, respectively. Perfect communication between the two is achieved even if proxies have been compiled into different languages or run on different implementations of ORB (e.g., *Rake Receiver* may be a unit of software running on a reconfigurable processor, and *Transmitter* is another unit of software running on a DSP).

7.3.3.2   CORBA Performance

The description of CORBA and Figure 7.15 show that there is some over-head in using the method, since interobject communications are not direct and must flow through at least one ORB. Performance issues were recognized by the SDRF, and the organization undertook some performance measurement prior to CORBA's inclusion in the DOCSRA. These measurements looked at the data flowing up and down the CORBA part of the stack and concluded that less than 20% of the processing time was spent in the CORBA part of the stack [4].

The SDRF also found that timing performance depends upon the location of the objects and the capabilities of the particular ORB. They found that collocated objects, where each object is within the same process or address space, provided the best results. The latency for communications was only 20 μs [4] when using a 200-MHz Pentium processor. The time-consuming process of marshaling and unmarshaling data is not required for the collocated case, and this accounts for the result. The worst timing performance was measured when the objects were on separate processors communicating with a GIOP/IIOP-compliant protocol. While performing a bidirectional 64 bytes of data, 420 μs were required between a Sun UltraSPARC 5 with a 270-MHz UltraSPARC IIi.

A "Real-Time CORBA Trade Study" [16] was undertaken by Boeing for the U.S. Defense Department to benchmark the following CORBA implementations: HARDPack from Lockheed Martin Federal Systems, ORBExpress RT from Objective Interface Systems, and The ACE ORB (TAO) from Washington University and Objective Computing, Inc. The study included a survey of users to gain additional statistics, including aspects of usability and documentation.

For the study Boeing ran the tests on a Motorola PowerPC VME-based processor running version 3 of the LynxOS operating system and a SPARC-based machine running version 2.6 of the Solaris operating system. Each test consisted of seven types of IDL operations with nine scenarios. Scenario 3a has the client and server objects on the one PowerPC device and deals with call and return (CR) operation tests plus one-way (OW) operation tests. Four sets of operations were reported for CR and OW: float operations, aligned records, nonaligned (NA) records, and CORBA Any transfers. Data sizes involved in the transfers ranged from 144 to 24,016 bytes. Results were only published [16] for TAO and ORBExpress, and an example is shown in Table 7.1; results are in milliseconds, and some results were removed for clarity (*).

Table 7.1
Call and Return Average Operation Times (in ms)

Operation	CORBA ORB	Data Size (bytes) >		
		144	12,016	24,016
Float	TAO	1.15	2.32	3.45
Record	TAO	1.16	3.09	5.12
NA Record	TAO	1.17	4.54	*
Float	ORBExpress	0.15	0.79	1.48
Record	ORBExpress	0.27	0.99	1.89
NA Record	ORBExpress	0.29	1.89	3.68

A major tradeoff for using the standardized method of communications that CORBA offers is that the amount (SLOC) of new communications and infrastructure software needed to be developed is reduced and made easier.

7.3.3.3 OMG IDL

IDL is part of the CORBA specification (Chapter 3). As illustrated in Figure 7.15, IDL is used to define the interface that sits on the outside of an object's boundary and to control how the object communicates with the outside world. This approach uses the principle of encapsulation, where the internal structure and mechanism of an object are kept inside a boundary where the client (e.g., ORB) is not allowed to penetrate. IDL allows the definition of interfaces that both client and server objects understand and can use regardless of platform, operating system, programming language, and so on. These interfaces specify the allowable operations and the associated input and output parameters with their "types," so that client and server can encode and decode values for travel over the communications medium. IDL enforces object orientation and strong exception handling to reduce the problems associated with incorrect invocations.

The following IDL interface is a very simple example. The object's "type" is transmitter, and it can perform two operations: setPilot and setTrafficChPwrInc. When increasing the traffic channel power for a given user (setTrafficChPwrInc), the transmitter accepts an unsigned integer "walshcode" specifying for which channel the function must be performed. The return value for setTrafficChPwrInc does not need a name and is also an unsigned integer; this may be used to indicate the success of the operation.

```
interface transmitter {
unsigned_int setPilot ();
unsigned_int setTrafficChPwrInc(in unsigned_int
walshcode);
}
```

7.3.4 Software Design Patterns

Software development has sufficient history behind it so that each new software project is almost certain to include design requirements and problems that have occurred in at least one preceding project. The advent of object-oriented programming practices has propelled the volume of software being developed, and this has helped to foster efforts to recognize common problems and publish well-designed solutions for them. These solutions are known as "design patterns," and the book *Design Patterns Elements of Reusable Object-Oriented Software* [17] details the four elements of a pattern: name, problem, solution, and consequences. Reference [17] details 23 patterns, including factory, singleton, proxy, builder, visitor, and others.

7.4 Component Choices

The first part of this chapter has considered possibilities for an SDR software architecture. These architectures are abstract and by their nature are independent of operating system and language used to develop the application software. This section briefly covers a range of operating systems and languages that can be selected during the detailed design phase for use during implementation.

7.4.1 Real-Time Operating Systems

The choice of an operating system will be driven by many requirements, including support for the target processor, real-time features, development tools, and cost (development and production).

7.4.1.1   LINUX & RT LINUX

LINUX [18] is a free UNIX type of operating system that is made available under the GNU general public license [19]. It was originally developed by Linus Torvlads in 1991, when it was released as version 0.02. The licensing arrangement ensures that users can obtain source code for the operating system. Once a user has the source code, he or she is permitted to create new versions, which they can charge for, but the new source code must be made available back to the community. The open source, low-cost environment has spurred on thousands of developers worldwide to contribute to LINUX, with the result that the operating system is now very widespread and increasing in popularity at an exponential rate.

While LINUX has been particularly popular in the PC world, it is generally not suitable for hard real-time embedded systems. Standard LINUX takes up to 600 μsec to start a handler and can be more than 20 μsec late for a periodic task [20]. The LINUX operating system is optimized for the general case and has some fundamental features that contradict real-time requirements. For example, LINUX will not preempt the execution of the lowest-priority tasks during system calls, synchronization is not fine enough and causes long periods when data is tied up by a non-real-time thread and unavailable to real-time threads, and LINUX will make high-priority tasks wait for low-priority tasks to release resources.

Real-Time LINUX (RT LINUX) has been developed to solve the shortcomings of standard LINUX and is now available to meet the growing need for a low-cost, real-time embedded operating system. RT LINUX treats the LINUX kernel as a task executing under a small, real-time operating system. The design has LINUX as the "idle" task for the real-time OS and only executing when there are no real-time tasks to run. In this mode the LINUX task cannot prevent itself from being preempted or allow the blocking of interrupts. This is achieved by the RT LINUX kernel intercepting LINUX requests to disable interrupts and recording them and then returning to LINUX. If there is a handler for the real-time event, it will be invoked. This ensures that LINUX cannot add latency to the real-time interrupt response time no matter what state LINUX is in.

RT LINUX is distributed by a commercial organization (FSMLabs), which was founded by the original product creators.

7.4.1.2   VxWorks

The VxWorks real-time operating system (RTOS) is a commercially available product from WindRiver. The operating system is a closed proprietary offering; however, it has grown in popularity and was selected for the high-profile 1997 Mars Pathfinder Lander project. WindRiver recently took over the pSOS operating system, and it is widely expected that VxWorks and pSOS will be merged in the future into a single product.

The operating system is a good choice for systems that only use general-purpose microprocessors (e.g., PowerPC, Intel Pentium, ARM, SPARC, MIPS, and so on). Major disadvantages for VxWorks when considering software radio are that the operating system does not support any digital signal processing devices (e.g., TI or Analog Devices), and it imposes a royalty fee for every deployed instance of the RTOS on a processor.

Features include unlimited multitasking, preemptive scheduling, round-robin scheduling, 256 priority levels, POSIX 1003.1 compatibility, and a good range of diagnostic tools. For larger, embedded systems there is support for networking protocols such as TCP/IP, PPP, FTP, SNTP, and others.

7.4.1.3 OSE

The OSE RTOS by Enea is similar to VxWorks in that it is also a closed proprietary and commercially available system with a similar licensing structure (royalty per instance). The advantages of OSE is that it supports both general-purpose microprocessors and DSPs and is certified for use in systems requiring a level of safety integrity [21].

7.4.1.4 MQX

A commercially available product with a different supply model is the MQX RTOS by Precise. The supplier has chosen a halfway house between LINUX and VxWorks by supplying the source code for MQX and making it royalty free. The RTOS suits mixed processor environments and is available for RISC and CISC microprocessors (e.g., PowerPC, ARC, ARM, MIPS) and DSPs (TIC6x, TIC5x, TIC4x, TIC3x, and ADSP2106x). By being provided with the source code for the operating system, the user has the choice to port to other processors if required.

MQX presents the user with a standard API regardless of the processor used.

7.4.1.5 DSP/BIOS

DSP/BIOS is a kernel provided by Texas Instruments that provides real-time operating support for DSPs. The kernel supports preemptive multitasking and other services that enable applications to more effectively use event-

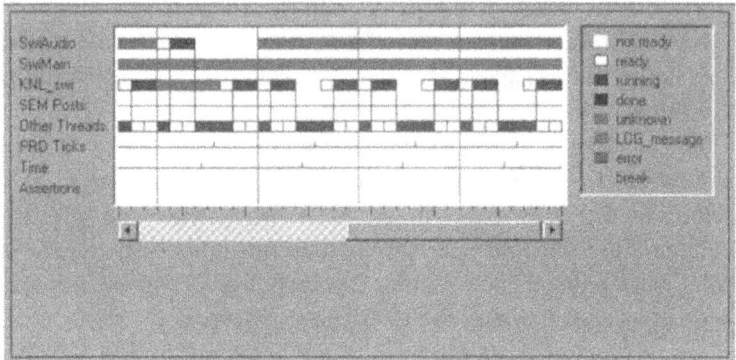

Figure 7.16 Example software and hardware tasks in DSP/BIOS.

driven and interrupt service paradigms. Application software can take advantage of traditional multitasking services, mailboxes, queues, semaphores, and resource protection locks.

Developers have the flexibility of selecting several I/O mechanisms, including data pipes and data stream models. The configuration of the kernel object programming model can be static or dynamic, and the kernel allows run-time memory management, providing dynamic memory allocations and deallocations. Application management and configuration of resources can be performed dynamically, thereby enabling developers to build self-configuring and more complex applications whose mix of functions changes over time.

Figure 7.16 illustrates the results of a running DSP/BIOS program as viewed through the Code Composer execution graph. Event resolution was set to one-millisecond intervals, and the figure shows the operating system switching between the kernel task (KNL_swi, a do nothing idle function), the SwiAudio function (software interrupt controlled), and Other Threads (includes hardware interrupts).

7.4.2 High-Level Software Languages

There are many high-level software languages; however, the most widely used in commercial software defined radio implementations are C and C++.

7.4.2.1 C Code for DSP

Dennis Ritchie of AT&T Bell Laboratories invented the C language in 1972 as a means of implementing the UNIX operating system in a high-level language. The design of C and its standardization by ANSI has led to a mass adoption of the language and a high degree of portability across many processors. Therefore, C was an obvious choice when DSP manufacturers first upgraded their assembly language compilers to accommodate a high-level language.

The following simple two-line C program loops through a series of additions, where the loop counter "i" is added to the "result" for each pass through the loop.

```
for (i = 0; i < n; i++) {
    result = result+i;
}
```

This C program can be written by almost any programmer in a very short period and compiled to run on a general-purpose Intel type processor or a DSP. An example of the assembly code produced by the Code Composer Studio compiler for the Texas Instruments' C62 DSP follows:

```
        ZERO.D2         B4
        STW.D2T2        B4,*+DP[0x11C]
        LDW.D2T2        *+DP[0x11B],B4
        LDW.D2T2        *+DP[0x11C],B5
        NOP             4
        CMPLT.L2        B5,B4,B0
[!B0]   B.S1            L2
        NOP             5
        L1:
        LDW.D2T2        *+DP[0x11A],B4
        LDW.D2T2        *+DP[0x11C],B5
        NOP             4
        ADD.D2          B5,B4,B4
        STW.D2T2        B4,*+DP[0x11A]
        NOP             2
        LDW.D2T2        *+DP[0x11B],B5
        ADD.S2          1,B5,B4
        STW.D2T2        B4,*+DP[0x11C]
        NOP             3
        CMPLT.L2        B4,B5,B0
[ B0]   B.S1            L1
        NOP             5
        L2:
        ZERO.D1         A4
        B.S2            B3
        NOP             5
```

The high-level language capability of C abstracts the program to such a point that it becomes independent of the target processor that will perform the execution. It becomes the job of the compiler to decide on which hardware-specific instructions to use, and this example illustrates that the result can be a significantly large number (26 in the example) of assembly language mnemonics.

In the past, most mass-manufactured radio products with a DSP software component have used assembly language to achieve the lowest cycle count and highest performance. Today's DSP C compilers are approaching efficiencies obtained by directly written assembly.

7.4.2.2   C++

C++ is an extension of the C language developed in 1983 by Bjarne Stroustrup of Bell Labs. The language was designed for UNIX and aimed to make programming easier and code more portable. It is an OO language and provides full inheritance, polymorphism, and data binding features. A major distinction between OO and structured languages such as C is that data and functions (operations on data) can be combined to create objects.

The language is starting to find its way onto DSP; however, at this stage C remains the dominant DSP language, particularly when code size is an issue. C++ is, however, an excellent choice for implementing the higher-level layers two and three protocol software in a cellular mobile phone or base station.

7.4.3   Hardware Languages

The hardware languages of VHDL and Verilog are widely used to program devices such as FPGAs and RCPs. Both languages specify functionality at various levels of abstraction from behavioral (most abstract) to gate level (most detailed). These languages can, however, be bypassed in favor of programming at the register transfer logic (RTL) level, where the functionality between individual storage elements (or registers) and timing is specified.

These hardware languages continue to be extended (e.g., Verilog 2001) and provide more abstraction and flexibility; however, they may be overtaken by recent efforts to target much higher level languages such as Java to FPGAs. Xilinx now provides Java support for the Virtex II range of FPGAs; the "LavaCore" instruction set implements the Java virtual machine instruction set, as specified in "The Java Virtual Machine Specification" by Lindholm and Yellin.

7.4.3.1 VHDL

VHDL, or VHSIC Hardware Description Language, has been around for many years and was developed as a consequence of the U.S. government program named Very High Speed Integrated Circuits (VHSIC). Conclusions from this program led to development of VHDL, a language for describing the function (or behavior) and structure of integrated circuits. The language is now standardized and available from the IEEE [22].

Complicated behavior in digital circuits cannot always be described by way of structures, inputs, outputs, and Boolean equations. These static methods cannot adequately deal with timing dependencies, and for this reason VHDL is designed as an executable language with similarities to the Ada high-level software language. VHDL is not as fully featured as Ada but includes a range of features, including identifiers, comments, literal numbers, strings, and data types such as integer and so on.

The following VHDL code illustrates a simple "for" loop example:

```
Architecture A of
CONV_INT is
begin
     process(VECTOR)
          variable TMP: integer;
     begin
          TMP := 0;
          for I in 7 downto 0 loop
               if(VECTOR(I)='1')
          then
                    TMP := TMP + 2**I;
               endif;
          endloop;

          RESULT <= TMP;
          end process;
     end A;
```

7.4.3.2 Verilog

Verilog is also a hardware description language (HDL); it predated VHDL and started out in 1985 as a proprietary language. Competition from the open VHDL language then forced Verilog down the standardization path and it too is available from the IEEE [23]. If VHDL is similar to Ada, then Verilog has structure akin to the C high-level language.

The following Verilog code illustrates a simple "for" loop example:

```
integer list [31:0];
integer index;

initial begin
    for(index = 0; index < 32; index = index +1)
        list[index] = index + index;
    end
```

7.5 Conclusion

This chapter has covered emerging software radio standards and the technologies being used to specify and support them. We have proposed expanding the software radio definition to include the use of object-oriented methodologies. The extent to which OO can be pushed into the core of a real-time software radio will rely on trading off performance and efficiency; however, this can be a staged process providing the software architecture supports it. The scope of the chapter has only allowed a cursory review of a range of software and hardware languages; there are many excellent books available for readers entering the detailed design and implementation stage of a software radio project.

References

[1] ETSI, "Base Station Controller—Base Transceiver Station (BSC-BTS) Interface: General Aspects," GSM 08.51, August 1999.

[2] Lynes, D., "Cellular BTS SDR Framework," *2001 International Conference on Third-Generation Wireless and Beyond*, San Francisco, CA, May 30, 2001.

[3] Joint Tactical Radio System (JTRS) Joint Program Office, "Software Communications Architecture Specification MSRC-5000SCA V2.2," November 17, 2001.

[4] Software Defined Radio Forum, "Distributed Object Computing Software Radio Architecture v1.1," July 2, 1999.

[5] IEEE, "IEEE Recommended Practice for Architectural Description of Software-Intensive Systems IEEE Std 1471-2000," October 9, 2000.

[6] Object Management Group, "OMG Unified Modeling Language Specification." Version 1.3, March 2000.

[7] Mercury Computer Systems, "AdapDev SDR Brochure," DS-5C-30, 2000.

[8] IEEE, "1003.13-1998 IEEE Standard for Information Technology—Standardized Application Environment Profile (AEP)—POSIX Real-time Application Support," 1998.

[9] Obenland, K., "The Use of POSIX in Real-Time Systems, Assessing Its Effectiveness and Performance," http://www.mitre.org/support/papers/tech_papers99_00/obenland_posix/obenland_posix.pdf, 2000.

[10] "Programmable Modular Communication System (PMCS) Guidance Document," July 31, 1997.

[11] International Organization for Standardization, "ISO/IEC 14750:1999 Information Technology—Open Distributed Processing—Interface Definition Language," 1999.

[12] Object Management Group, "The Common Object Request Broker: Architecture and Specification CORBA 2.4.2," February 2001.

[13] Software Defined Radio Forum, "Distributed Object Computing Software Radio Architecture Vol. 1.1, July 2, 1999, pp. 2–11.

[14] Fowler, M., and K. Scott, *UML Distilled Applying the Standard Object Modeling Language*, Reading, MA: Addison-Wesley, 1997, p. 50.

[15] http://www.omg.org.

[16] Boeing, "Real-Time CORBA Trade Study," D204-31159-1, January 10, 2000.

[17] Gamma, E., R. Helm, R. Johnson, and J. Vlissides, *Design Patterns—Elements of Reusable Object-Oriented Software*, Reading, MA: Addison-Wesley, 1998.

[18] http://www.linux.org.

[19] http://www.gnu.org/licenses/gpl.html.

[20] Yodaiken, V., "The RT LINUX Manifesto," Department of Computer Science, New Mexico Institute of Technology, http://www.fsmlabs.com/developers/white_papers/rtmanifesto.pdf.

[21] Enea OSE Systems Inc., "Overview of the IEC 61508 Certification of the OSE RTOS," R1.0, 1999.

[22] IEEE, "VHDL IEEE Std 1076-1993," 1993.

[23] IEEE, "Verilog IEEE Std 1364-1995," 1995.

8

Applications for Wireless Systems

The 3G air interfaces are recognized as being significantly more complicated than those used for 2G cellular systems. The degree of this complexity has a significant effect on the choice of signal processing components, as suggested in Chapter 6 (e.g., Table 6.1), where we estimated the processing requirements for a UMTS/WCDMA transmitter and receiver. Earlier chapters have reviewed chip-rate and symbol-rate processing, so in this chapter we provide additional details of the layer one air interface specifications for the leading 3G technologies (UMTS/WCDMA, CDMA2000, and EDGE). These specifications are administered by 3GPP (http://www.3gpp.org) and 3GPP2 (http://www.3gpp2.org); they do not mandate any particular implementation.

8.1 Introduction

This chapter presents examples of software defined radio implementations in the mobile cellular domain. Some of these SDRs are commercial products, and others are research-based testbeds. SDRs are now being made available in the marketplace both as black boxes (e.g., BTS/Node B) and as generic SDR platforms suitable for applications development in the 3G wireless domain.

This chapter concludes by reviewing an early 3G CDMA2000 network in Korea and discusses the configurations of the base stations and other network components.

8.2 3G Air Interface Fundamentals

UMTS/WCDMA (3GPP) and CDMA2000 (3GPP2) are the 3G systems generating the most interest and activity; WCDMA has a FDD and a TDD mode—references in this chapter to WCDMA assume the FDD mode.

The WCDMA and CDMA2000 air interfaces use variants of CDMA; this section starts by covering the common fundamental principles behind the technology and then follows with details of the specifics for WCDMA and CDMA2000, respectively. The final part of the section introduces the TDMA-based GSM EDGE system as a potential 3G technology. GSM is in use by more people in the world today than any other system, and some carriers are proposing deploying the GSM EDGE enhancement either as their high data rate solution or as a stepping stone to WCDMA.

8.2.1 CDMA Fundamentals

Table 8.1 lists the major similarities and differences between WCDMA and CDMA2000.

WCDMA uses a completely new air interface, whereas CDMA2000 has evolved from IS-95. The two systems, however, share the core CDMA technology with some fundamental differences: systemwide BTS synchronization and incompatible modulation and spreading schemes.

8.2.1.1 Spreading

WCDMA and CDMA2000 use direct sequence spread spectrum to transmit and receive digital information over the air interface. Another method of

Table 8.1
Major 3G CDMA and WCDMA Parameters

	WCDMA	CDMA2000
Carrier bandwidth	5 MHz	1.25 MHz
Chip rate	3.84 Mcps	1.2288 Mcps
Power control frequency	1,500 Hz up-/downlink	800 Hz up-/downlink
Base station synchronization	Not applicable	Yes, usually GPS
Downlink transmit diversity	Yes	Yes
2G legacy support	GSM at the network level	IS-95 at the air interface

spreading called fast (relative to the symbol rate) frequency hopping is also used in radio systems to produce a similar increase in occupied spectrum. Figure 8.1 provides a simple illustration of direct sequence spreading.

The sequence of symbol bits {100100} is shown with its corresponding bipolar (1, −1) time domain signal waveform. For this case the spreading code is the series of four binary bits that occur during a symbol period (i.e., {1100}); this is also expressed as a four-chip sequence. The on-air (spread) signal is produced by multiplying the symbol signal by the spreading signal. This example uses a spreading factor (SF) of four and increases the transmitted bandwidth by the same factor. The on-air bits are modulated onto an RF carrier for transmission to the receiver, where the reciprocal process is applied. After demodulation in the receiver and synchronization with the spreading code, the on-air bits are multiplied and integrated to recover the transmitted symbols. Considering the first four chips, the receiver calculates $(1 \times 1 + 1 \times 1 + -1 \times -1 + -1 \times -1 = 4)$ and recovers the first symbol as a binary one.

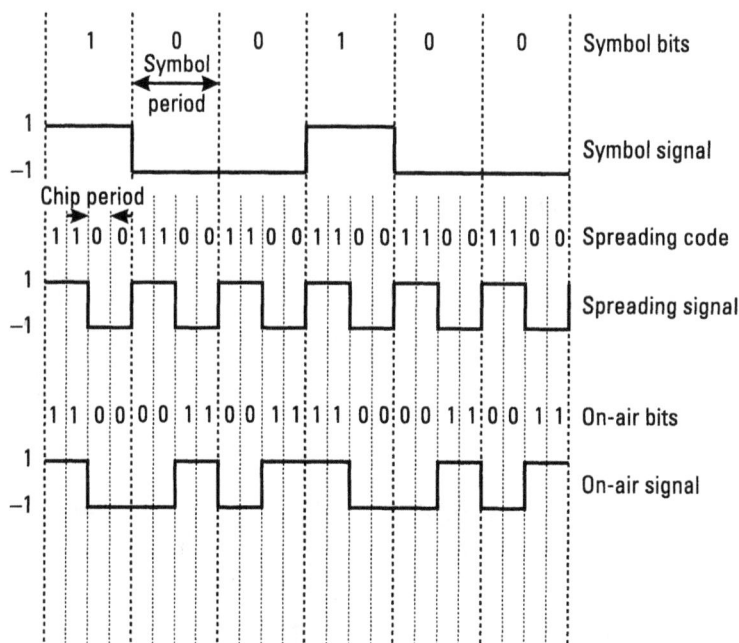

Figure 8.1 Direct sequence spreading.

8.2.1.2 Multiple Access

Spreading codes are also used to implement the multiple access part of
CD(MA) by choosing groups of codes that exhibit good orthogonal properties between each other. Two periodic functions (e.g., c_1 and c_2) are said to be
orthogonal when the result of multiplying and integrating them over their
period (T) equates to zero:

$$\int_0^T c_1(t).c_2(t)\,dt = 0 \tag{8.1}$$

Orthogonality allows different information streams to be multiplexed
(added) together for transmission through a single physical pipe (e.g., radio
channel) and demultiplexed without loss at the other end. Figure 8.2
illustrates the principle by considering the four-chip spreading code 1 (1100)
from Figure 8.1 and an orthogonal spreading code 2 (1010). The progressive
correlation (integration of the product) is plotted for despreading of the on-

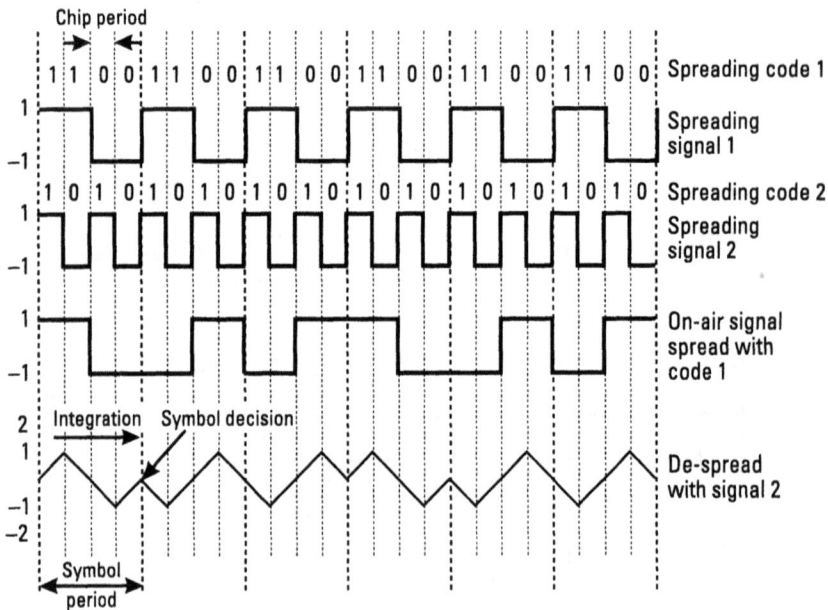

Figure 8.2   Orthogonal codes.

air signal (spread with code 1) by using code 2. The result at the end of each symbol period is always zero, indicating that codes 1 and 2 are completely orthogonal. In a real system with noise, unmatched amplitude, and imperfect timing the result will be a random variable (with mean and variance) with a finite probability that decision algorithms will produce an error.

8.2.1.3 Rake Receiver

Mobile communication is generally conducted in situations where the transmission channel is far from the ideal of a simple AWGN channel. An AWGN channel is possible for deep space communications, but in an urban environment, where the mobile is located anywhere within an array of objects, including human beings, buildings, and vehicles, the channel is usually characterized by several strong reflections or multipath signals. Considering the case of transmission between a mobile and a base station, each of the multipath signals will usually arrive at the receiver with different time offsets; in some cases the direct wave will be below the receiver's detection threshold and the receiver may only be able to recover reflected signals.

Multipath is a significant problem in time division multiple access systems such as GSM; it causes intersymbol interference (as does cochannel interference) and is usually minimized by using an equalizer (time domain filter). One of the great advantages with direct sequence spread spectrum transmission is that the multipath signals can be used to an advantage, and in some respects the system performance relies on it.

Figure 8.3 illustrates a four-finger Rake receiver, as typically deployed in a base station. Each finger of the rake receives oversampled (2X, 4X, 6X, or 8X typically) baseband in-phase (I) and quadrature (Q) digital data from a digital downconverter; this time series data is also fed to the path searcher.

In sequence of operation the path searcher despreads the digital downconverter output data; it completes this operation many times using different time delays (or code offsets). The range of time searched equates to the radius of the cell plus an allowance for delay spread. Strong multipath signals will appear as peaks, as illustrated in the path searcher block in Figure 8.3. The searcher sets a peak detection threshold to avoid false triggering and selects the best four peaks. The associated time delays for each of the peaks are forwarded to the fingers' fine finger tracking, code generator, and delay equalizer. The path searcher is usually computationally more intensive than the Rake because of this multiple search process; however, the signal processing load can be reduced by shortening/simplifying the despreading code and/

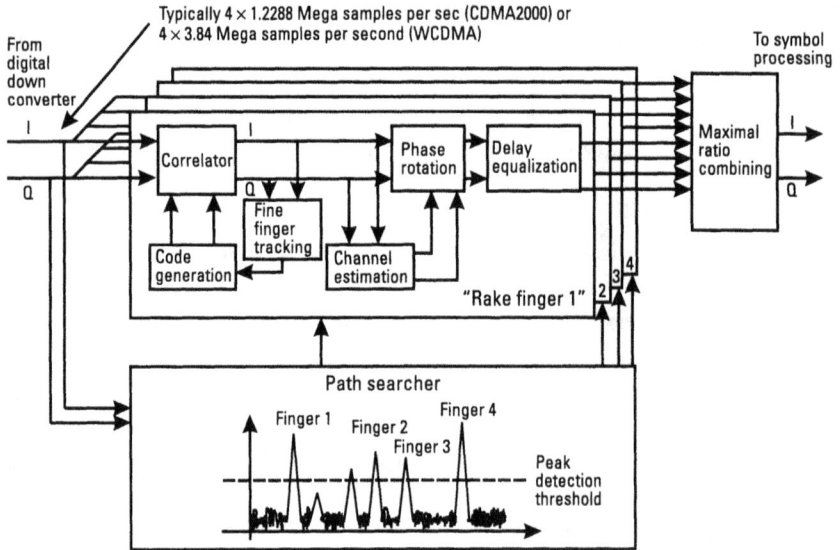

Figure 8.3   Rake receiver.

or reducing the calculation's duty cycle (code offset update rate). Simplifying the code may reduce the probability of detecting a suitable multipath, and reducing the update rate when the mobile is moving will tend to reduce the average signal to noise ratio for a given finger.

The timing information received by the code generation function is used to synchronize the despreading code and allow the finger to lock onto an individual multipath. The correlator (or despreader) multiplies the incoming data by the despreading code and integrates it (i.e., the inverse of Figure 8.1) to produce symbols. However, the timing correction is limited to the oversampling rate at which the data is processed (e.g., one-fourth chip for 4X oversampling), and the resultant symbol data can still have some phase rotation. This is removed by the phase rotation function, it uses phase data from the channel estimator, which, in turn, uses the pilot symbols for estimating the channel state.

Because the time delay for the path can change in-between updates from the path searcher, the finger needs to track any small path changes to ensure that an adequate signal to noise ratio is maintained. It does this with the feedback loop controlled by the fine finger tracking function. Code generation and the correlator periodically produce I and Q values on one-fourth chip either side (the early and late values) of the code value as assigned by the path searcher. Fine finger tracking measures the early and late correlations to check if the path is becoming shorter or longer and corrects the code generator as required.

The final step in the process is for the combiner to add the active outputs (inactive fingers = insufficient multipaths) with the aim of producing a larger signal to noise ratio when compared with that of a single finger. Because the propagation characteristics for each multipath will most probably be different, combination by simple addition may not increase the signal to noise ratio. Figure 8.3 considers a maximal ratio combiner; this algorithm weights the contribution of each finger to ensure that the result maximizes the signal to noise ratio.

Rake receiver implementations can vary for different standards (IS-95, UMTS, and CDMA2000) and for up- and downlinks; however, the principles for direct sequence spread spectrum systems remain similar for each of these options.

8.2.1.4 Soft Handover

In contrast to the TDMA technique of hard handover, all mobile CDMA systems use soft handover in an effort to reduce the dropped call rate during a mobile's transitions between cells or sectors. To achieve soft handover the mobile periodically searches for candidate base stations and relays their power strength measurements back to the current base station. The base station or base station controller uses an algorithm to control the handover process. Before breaking the circuit with the current base station or sector within a base station, the base station controller sets up the new target circuit and then, if the correct set of conditions is met (usually power level and time), the current circuit is dropped.

8.2.1.5 Power Control

All multiple access spread spectrum systems must use some form of power control mechanism to ensure adequate capacity. As indicated in the orthogonal example in Figure 8.2, spread signals from different users/channels must be closely matched in amplitude to maintain the degree of orthogonality and limit the degree of interference. WCDMA and CDMA2000 control the power of the up- and downlink transmissions at a rate that balances increased system capacity with increased complexity.

8.2.2 WCDMA

In response to the ITU's request for IMT-2000-compliant 3G systems (ITU-R
M.1225 [1]), the European Telecommunications Standards Institute (ETSI)
proposed the UMTS Terrestrial Radio Access (UTRA) RTT candidate submission [2] in January 1998. UTRA uses the WCDMA air interface for communications between the mobile and the network.

Interim WCDMA standards are developed by the Third-Generation
Partnership Project (3GPP) and then formalized as published documents by
ETSI. The current 3GPP organizational partners [3] include ARIB, CWTS,
ETSI, T1, TTA, and TTC.

WCDMA FDD can be divided into the physical layer (layer one) and
the upper layers (layers two and three); this section covers the fundamental
concepts of the physical layer one only, since it is the most relevant to the
software radio issues covered in this book (see Figure 8.4 [4]). The physical
layer one is generally a pipe for data bits and follows the principles of layered protocol stacks. Layer one only has the job of receiving bits from the
upper layers and sending them on; it has no need to know which bits are
content and which are control. The UMTS Terrestrial Radio Access Network (UTRAN) terminology for a mobile is a Ue and a base station is a
Node B.

Figure 8.5 illustrates a Node B view of the up- and downlink physical
layer processing flow and includes references to the appropriate 3GPP specification (e.g., spreading and modulation is covered by TS 25.213 [5]).

Multiple transport channels including user data and control messages
are processed and mapped to a number of physical channels. These physical

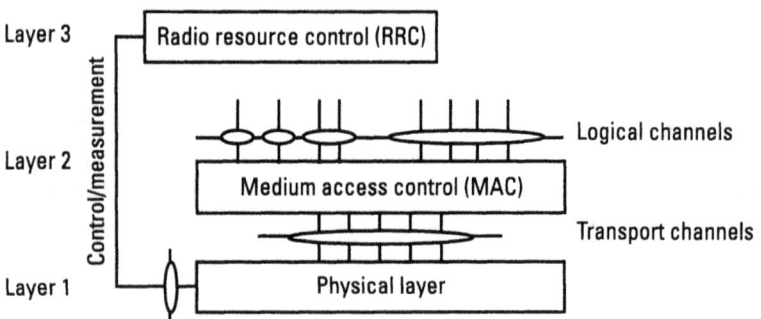

Figure 8.4   Interfaces with the physical layer. (©ETSI 2001. Further use, modification, or redistribution is strictly prohibited; ETSI standards are available from http://pda.etsi.org/pda/
and http://www.etsi.org/eds/. Reprinted with permission.)

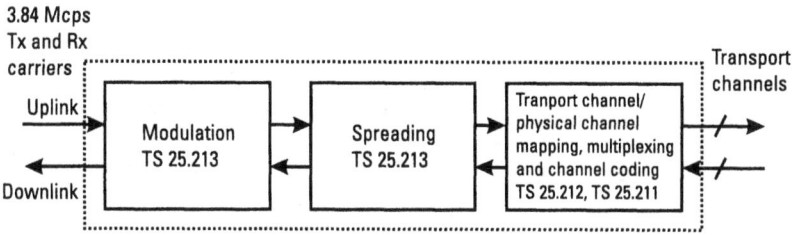

Figure 8.5 Physical layer processing blocks.

channels then go through a spreading stage, first by using a 3.84 Mcps code and then by a second scrambling code to enable cell identification for the downlink and for mobile/Ue identification on the uplink. The final stage results in a transmit and receive (for FDD) baseband 3.84 Mcps data stream, which is then modulated onto a radio carrier at the required frequency.

8.2.2.1 Transport Channels and Physical Channels

The higher-layer data is transmitted over the air using transport channels; these are mapped in the physical layer to different physical channels, as depicted in Figure 8.6 [6]. Physical channels are defined by a carrier frequency, scrambling code, channelization code (optional), time duration, and, on the uplink, relative phase (0 or $\pi/2$). Time durations are defined by start and stop instants, measured in integer multiples of chips. A radio frame corresponds to 38,400 chips and consists of 15 slots. Each slot consists of fields containing bits, and the length of a slot corresponds to 2,560 chips.

A major 3G advancement with WCDMA is the ability to support bandwidth on demand services; this is achieved by variable bit rate transport channels sent through the physical layer, with the system allowing multiple services to be multiplexed on the one circuit.

The transport format indicator (TFI) [7] is used in the interlayer communication between layer two and layer one each time a transport block set is exchanged between the two layers on a transport channel. Layer one combines TFIs from the group of transport channels it is responsible for sending to produce a transport format combination indicator (TFCI). The TFCI is appended to the physical control channel to indicate to the receiver the transport channels that are active for the current frame.

Transport channels are divided into dedicated channels and common channels. As the names suggest, dedicated channels are reserved for a single user, whereas common channels are shared among all the users in the serving cell.

Transport channels Physical channels

DCH ──────────── Dedicated physical data channel (DPDCH)
                 Dedicated physical control channel (DPCCH)
RACH ─────────── Physical random access channel (PRACH)
CPCH ─────────── Physical common packet channel (PCPCH)
                 Common pilot channel (CPICH)
BCH ──────────── Primary common control physical channel (P-CCPCH)
FACH ══════════▶ Secondary common control physical channel (S-CCPCH)
PCH ─────▶

                 Synchronisation channel (SCH)
DSCH ─────────── Physical downlink shared channel (PDSCH)
                 Acquisition indicator channel (AICH)
                 Access preamble acquisition indicator channel (AP-AICH)
                 Paging indicator channel (PICH)
                 CPCH status indicator channel (CSICH)
                 Collision-detection/channel-assignment
                 indicator channel (CD/CA-ICH)

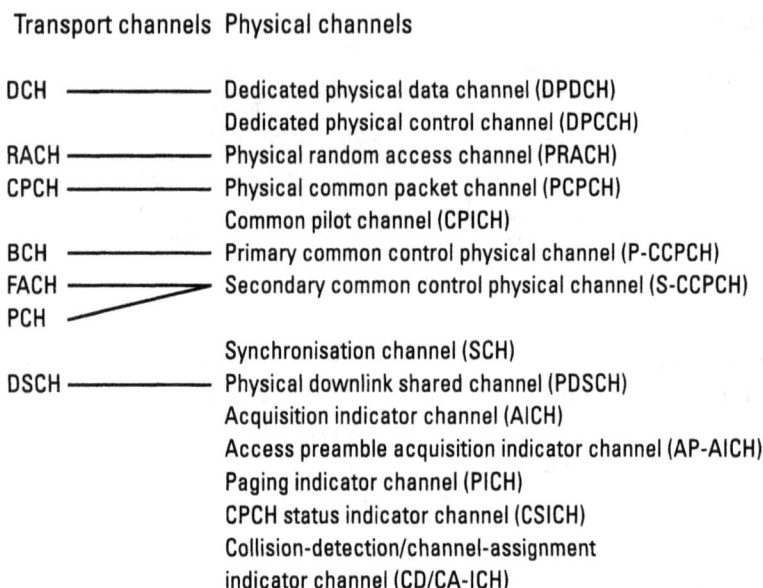

Figure 8.6   Transport channel to physical channel mapping. (© ETSI 2001. Further use, modification, or redistribution is strictly prohibited; ETSI standards are available from http://
pda.etsi.org/pda/ and http://www.etsi.org/eds/. Reprinted with permission.)

The DCH is the only dedicated transport channel and it carriers all the
service (e.g., speech frames) and control data (e.g., power control) from the
higher layers for the associated user. The DCH support fast (1,500 Hz)
power control, soft handover, and other features.

Figure 8.6 illustrates the six types of common transport channel:
RACH, CPCH, BCH, FACH, PCH, and DSCH.

The random access channel (RACH) carries control information in the
uplink (Ue to Node B) and is most often used by a mobile during initial
communications (logging on and call setup). Since the channel is shared by
all users in the cell, the data rate is very low and mobile transmissions last less
than two frames.

The common packet channel (CPCH) is also used in the uplink and
extends upon the RACH by carrying packet-based user data to the Node B.
It has additional features (compared with the RACH) that increase system
capacity (fast power control) and improve reliability (collision avoidance
mechanism). The packet data can extend over several frames.

The broadcast channel (BCH) is used on the downlink to send information specific about the network and the cell to all users in the cell. The

BCH is the first channel listened to by any mobile and includes information such as, the network code, random access codes, available random access slots, and the types of transmit diversity used in the cell.

The forward access channel (FACH) is used on the downlink to carry control information to users who have logged on to the cell. There can be more than one FACH, and they can also be used to carry packet data.

The paging channel (PCH) is used on the downlink to send control information to mobiles in the case of a network-originated call. The same messages can be sent to one or more cells based on the network's knowledge of where the mobile may be.

The last of the common channels is the downlink shared channel (DSCH). It is used to carry both user data and control information. It is similar to the FACH except that it uses fast power control, does not need to be broadcast over the whole cell, uses variable bit rate, and can use transmit diversity.

Transport channels are transmitted using 10-ms frames; for identification purposes a 12-bit system frame number (SFN) is used. System and control messages can span more than one 10-ms frame.

In Figure 8.6 we can see that some physical channel types (e.g., synchronization channels) do not map to any transport channel type and remain solely in the realm of layer one. This is because these channel types are only needed for proper functioning of the air interface (i.e., between the Node B and the mobile).

8.2.2.2 Synchronization Channel

The WCDMA system has been designed so that base stations do not have to be synchronized to a common time reference (e.g., GPS). Even if base stations are synchronized to each other, when a mobile is first switched on it will have no prior information regarding the cell's timing. The synchronization channel (SCH) is provided by base stations to allow mobiles to achieve slot and frame synchronization. There are two subchannels: the primary and secondary SCH. The 10-ms SCH frame is divided into 15 slots; each is 2,560 chips in length.

The primary SCH contains a 256-chip modulated code called the PSC; this one code is transmitted at the beginning of each slot by every base station in the network. At the same time as the PSC is transmitted, the secondary SCH transmits the secondary synchronization codes (SSCs). The SSCs are a sequence of 15 codes repeatedly transmitted over a 15-slot period, where each code is 256 chips in length. There are 16 different SSCs, which can be transmitted in any given slot, and these 16 codes are used to create 64 different (length 15) sequences. The sequence is used to tell the mobile to which one of 64 code groups the base station's downlink scrambling code belongs.

When first acquiring synchronization, the mobile searches (performs correlations) for the PSC over at least a frame boundary; the timing for the strongest peaks (should see a series of equally spaced ones) is used as the location of the slot boundaries. The mobile can then decode the secondary SCH and determine frame synchronization.

8.2.2.3 Dedicated Channel

The frame structure for the dedicated channel is depicted in Figure 8.7 [6].

The dedicated transport channel carries all the data and control information for a particular user, and, as the frame structure indicates, data (DPDCH) and control (DPCCH) are time multiplexed.

The DPDCH can support variable bit rate by one of two methods. The first method is for the case when the TFCI is not present, spreading rate remains fixed, and lower bit rates are achieved by discontinuous transmission. In the second method the TFCI is available, and bits are repeated while maintaining continuous transmission to reduce the effective throughput rate.

A user data rate of between 1 and 3 Kbps can be sustained with a spreading factor of 512; the maximum IMT2000 user data rate of 2.3 Mbps can be reached by reducing the spreading factor to four and by using three parallel codes. In this case, the system effectively reduces to one with no

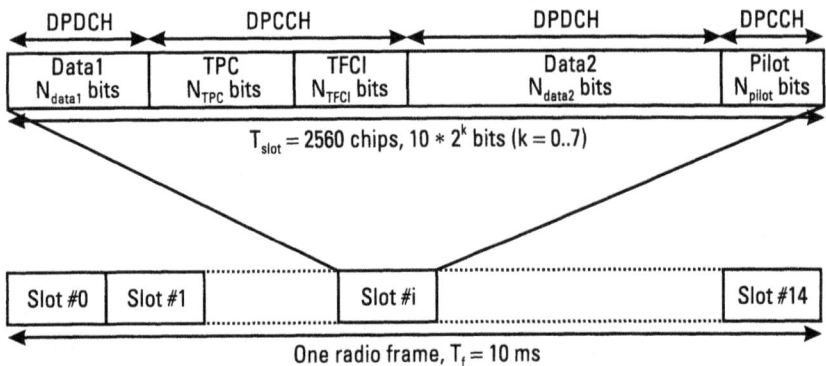

Figure 8.7 Downlink DPCH frame structure. (© ETSI 2001. Further use, modification, or redistribution is strictly prohibited; ETSI standards are available from http://pda.etsi.org/pda/ and http://www.etsi.org/eds/. Reprinted with permission.)

spreading or processing gain and requires a near perfect transmission channel to maintain a low bit error rate. A 2-Mbps user data will only be possible for stationary mobiles at very close range, with no other user interference present.

8.2.2.4 Multiplexing and Channel Coding

Transport channel multiplexing and channel coding follows the steps illustrated in Figure 8.8 [8]. First, cyclic redundancy check (CRC) bits are added to allow error detection on transport blocks. The transport blocks (TrBKs) are then either concatenated or segmented into different blocks to fit the available block size. Error correcting codes (convolutional or Turbo) are then added followed by rate matching; this process matches the number of bits for transmission with the number of available bits by using repetition or by puncturing. Bits are then interleaved to reduce the number of sequential errors during fading (makes error correction easier), and, finally, the different transport channels are serially multiplexed (TrCH multiplexing) together on a frame by frame basis. More than one spreading code can be used to segment the channel and provide more capacity.

8.2.2.5 Spreading

In the WCDMA system the variable symbol rate bandwidth (several kilobytes per second to 2.3 Mbps) is increased by applying a spreading (or channelization) code—hence, the term "(W)ideband," where the resultant radio carrier occupies 5 MHz of spectrum. The spreading codes exhibit good orthogonal properties between each other, and they are used to multiplex users on the downlink and for identification of terminals on the uplink. Figure 8.9 [5] illustrates the use of channelization codes where the complex valued downlink physical channel (except SCH) is split into real and imaginary parts and then multiplied by the code $C_{ch,SF,m}$ (SF = spreading factor and m = code number).

These codes use the orthogonal variable spreading factor (OVSF) technique and are generated from a single base Walsh-Hadamard matrix:

$$C_{2N} = \begin{bmatrix} C_N & C_N \\ C_N & -C_N \end{bmatrix} \qquad (8.2)$$

The OVSF codes can be viewed as a tree (see Figure 8.10 [5]).

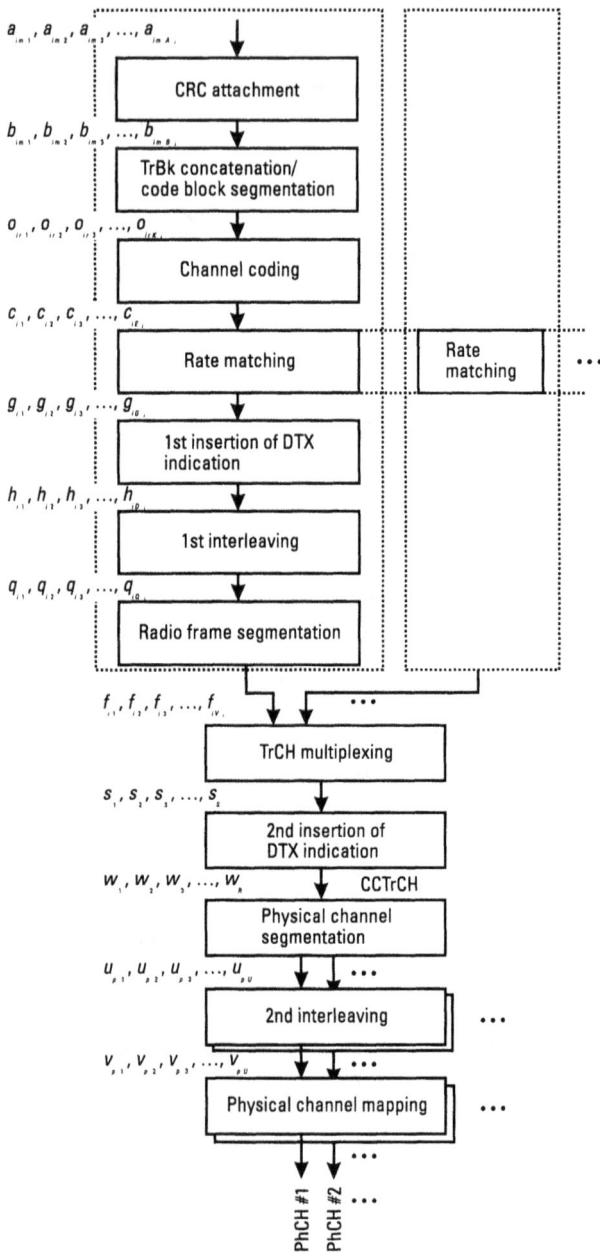

$a_{im1}, a_{im2}, a_{im3}, ..., a_{imA_i}$

CRC attachment

$b_{im1}, b_{im2}, b_{im3}, ..., b_{imB_i}$

TrBk concatenation/
code block segmentation

$o_{ir1}, o_{ir2}, o_{ir3}, ..., o_{irK_i}$

Channel coding

$c_{i1}, c_{i2}, c_{i3}, ..., c_{iE_i}$

Rate matching

Rate
matching

$\cdots$

$g_{i1}, g_{i2}, g_{i3}, ..., g_{iG_i}$

1st insertion of DTX
indication

$h_{i1}, h_{i2}, h_{i3}, ..., h_{iP_i}$

1st interleaving

$q_{i1}, q_{i2}, q_{i3}, ..., q_{iQ_i}$

Radio frame segmentation

$f_{i1}, f_{i2}, f_{i3}, ..., f_{iV_i}$ $\cdots$

TrCH multiplexing

$s_1, s_2, s_3, ..., s_S$

2nd insertion of
DTX indication

$w_1, w_2, w_3, ..., w_R$ CCTrCH

Physical channel
segmentation

$u_{p1}, u_{p2}, u_{p3}, ..., u_{pU}$ $\cdots$

2nd interleaving $\cdots$

$v_{p1}, v_{p2}, v_{p3}, ..., v_{pU}$ $\cdots$

Physical channel mapping $\cdots$

$\cdots$

PhCH #1 PhCH #2 $\cdots$

Figure 8.8 Downlink transport channel multiplexing. (© ETSI 2001. Further use, modification, or redistribution is strictly prohibited; ETSI standards are available from http://pda.etsi.org/pda/ and http://www.etsi.org/eds/. Reprinted with permission.)

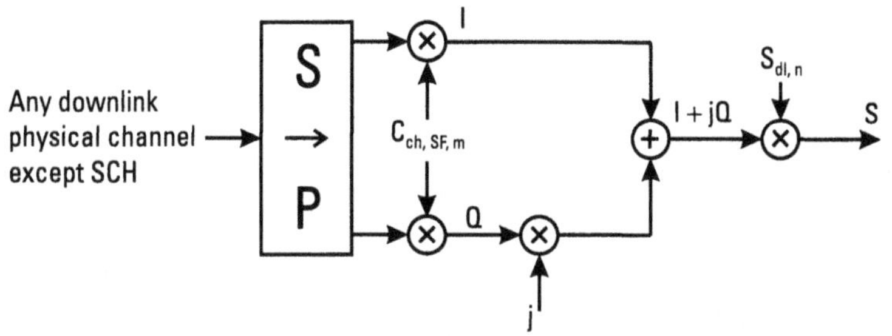

Figure 8.9 Spreading for downlink physical channels. (©ETSI 2001. Further use, modification, or redistribution is strictly prohibited; ETSI standards are available from http://pda.etsi.org/pda/ and http://www.etsi.org/eds/. Reprinted with permission.)

The third level of the tree has four adjacent branches, each containing a spreading factor four code. The codes on higher branches ($c_{ch,4,m}$ codes are on branches above $c_{ch,2,m}$ codes) are generated by the codes on the branches immediately below them, as illustrated by the code relationship formula in Figure 8.10; this is just a different representation of the single base Walsh-Hadamard matrix. Expressing the four $c_{ch,4,m}$ codes as a series of $c_{ch,2,m}$ codes

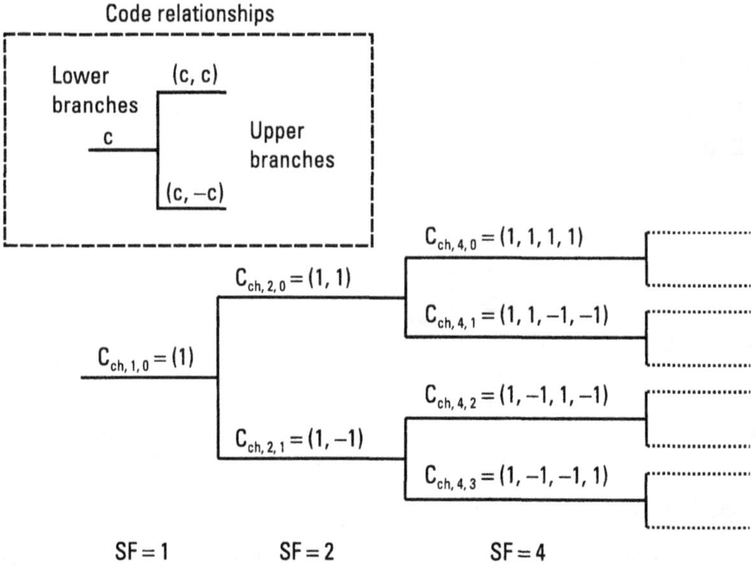

Figure 8.10 OVSF code tree. (© ETSI 2001. Further use, modification, or redistribution is strictly prohibited; ETSI standards are available from http://pda.etsi.org/pda/ and http://www.etsi.org/eds/. Reprinted with permission.)

we end up with the following, $\{c_{ch,4,0}, c_{ch,4,1}, c_{ch,4,2}, c_{ch,4,3}\} = \{(c_{ch,2,0}, c_{ch,2,0}),$ $(c_{ch,2,0}, -c_{ch,2,0}), (c_{ch,2,1}, c_{ch,2,1}), (c_{ch,2,1}, -c_{ch,2,1})\}$. This inheritance of codes from lower branches places restrictions on code use when mixing different spreading rate signals. A code with a given SF cannot be used with a code of a lower SF living on the same chain of branches either (e.g., $c_{ch,4,0}$ and $c_{ch,2,0}$ would not be used together in order to maintain orthogonality).

After spreading by a channelization code, WCDMA employs a subsequent level of spreading called scrambling. Figure 8.9 shows multiplication by the scrambling code $s_{dl,n}$ as the final stage. The chip rate of the signal is not changed during this final multiplication, and this extra code is used only for identification purposes. On the uplink it is used to separate users, and on the downlink it is used to separate base stations in the network.

8.2.2.6 Modulation

Following the spreading operations, WCDMA uses QPSK to modulate the complex valued chip sequences, both for the uplink and downlink. Figure 8.11 [5] shows the modulation process; ω is the carrier frequency and the pulse shaping functions preceding the I and Q multipliers is a root-raised cosine filter. This filter has a roll-off α of 0.22 in the frequency domain to ensure that the transmitted bandwidth stays within the specification.

8.2.3 CDMA2000

Synchronous CDMA started with the IS-95 (also known as CDMA One) standard developed by Qualcomm and the TIA. IS-95 was the world's first commercially based CDMA mobile system, with the IS-95A standard sup-

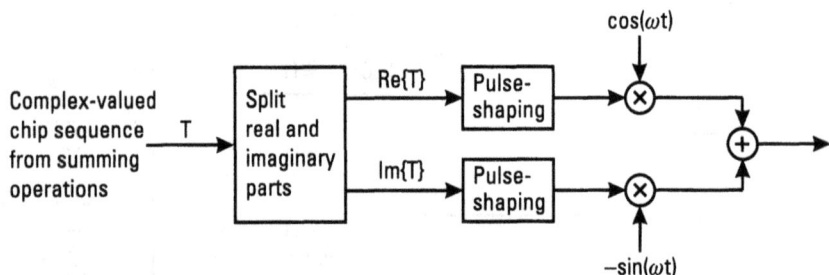

Figure 8.11 Modulation. (© ETSI 2001. Further use, modification, or redistribution is strictly prohibited; ETSI standards are available from http://pda.etsi.org/pda/ and http://www.etsi.org/eds/. Reprinted with permission.)

porting 9,600 bps traffic channels (Rate Set 1) and the enhanced IS-95B standard providing an increased data rate of 14,400 bps (Rate Set 2).

In response to the ITU's request for IMT-2000-compliant 3G systems (ITU-R M.1225 [1]), the TIA proposed the CDMA2000 system [9] in July 1998. CDMA2000 is an extension of IS-95 and has backward compatible modes that allow a smooth evolution from 2G to 3G. This CDMA2000 proposal was followed by the creation of the Third-Generation Partnership Project 2 (3GPP2) at the end of 1998 with the TIA, CWTS, TTA, TTC, and ARIB as members. Although there were efforts to join the 3GPP and 3GPP2, it was eventually agreed that the two projects would continue in parallel to develop a wideband (UMTS/WCDMA) solution and a narrowband multicarrier (CDMA2000) solution.

8.2.3.1 Physical Layer One

In principle CDMA2000 uses the same three (L1, L2, and L3) layered architecture as WCDMA. Figure 8.12 [10] shows layer one as the physical layer, layer two is split into a medium access communication (MAC) sublayer and a link access communication (LAC) sublayer, and layer three is known as the upper layer. This section restricts its coverage to layer one (see [11] for complete details), since it is the most relevant to the software radio issues covered in this book.

The CDMA2000 candidate RTT submission [9] proposed three spreading rates: SR1 at 1.2288 Mcps, SR3 at 3.6864 Mcps, and SR6 at 7.3728 Mcps; however, the current physical layer standard [11] has dropped SR6. SR1 provides backward compatibility for IS-95 RF carriers and occupies the same 1.25-MHz-wide radio channel; the term CDMA2000-1X is used to mean one carrier of SR1 in the forward and reverse links. SR3 is possible on the forward link by grouping three adjacent SR1 carriers and on the reverse link by direct spreading of a single radio carrier; this mode is also known as 3X and occupies 3.75 MHz of bandwidth. The "multicarrier" or "MC" terminology for CDMA2000 comes about as a result of the SR3 operation on the forward link. The standard handles asymmetric data throughput (e.g., Web surfing and file uploading) quite well by allowing SR3-MC on the downlink with SR1 on the uplink or SR1 on the downlink and SR3-MC on the uplink.

The CDMA2000 standard is defined by nine radio configurations (RCs), as summarized in Table 8.2 for the forward traffic channel.

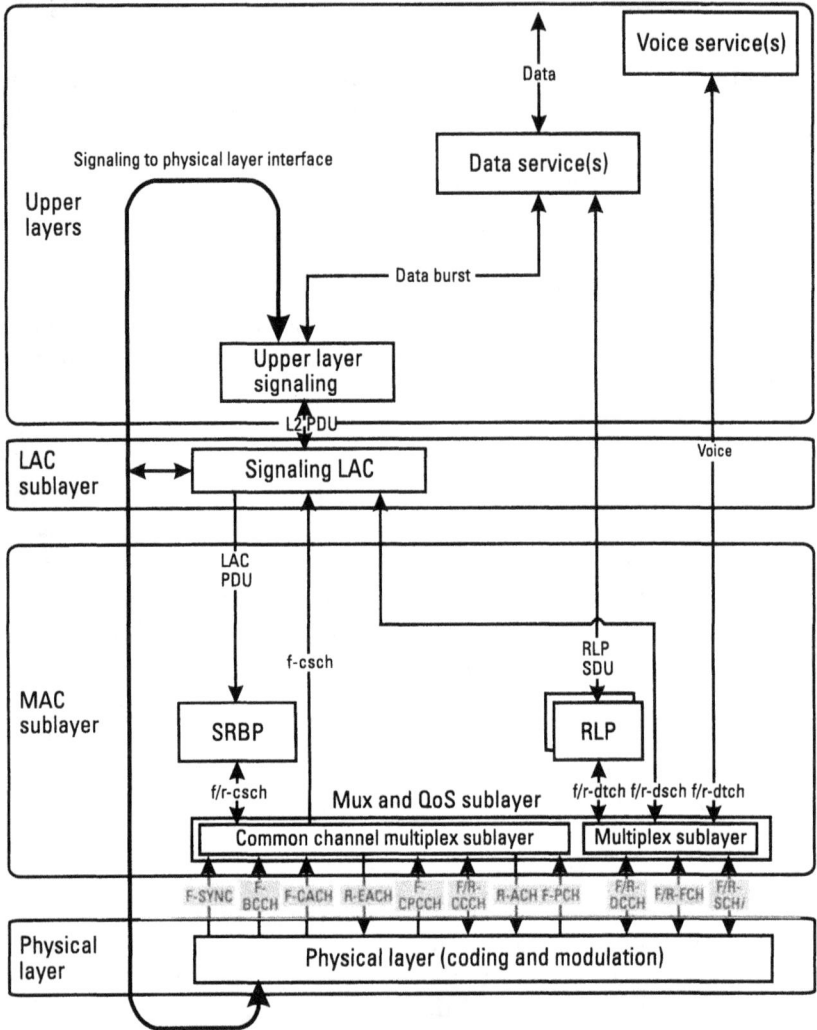

Figure 8.12 CDMA2000 architecture. (*Source:* 3GPP2, 2001. Reprinted with permission.)

8.2.3.2 Synchronous Operation

The CDMA2000 system maintains a synchronous architecture, as derived from the IS-95 standard; system time is referenced back to the origin of GPS time(i.e., January 6, 1980, 00:00:00 UTC).

Each base station has a code clock, which is the combination of the pseudonoise (PN) short code and long code generators used in the spreading

Table 8.2
Forward Traffic Channel Radio Configuration Characteristics

Radio Configuration	Associated Spreading Rate	Data Rates and Modulation Method
1	1	1.2–9.6 Kbps, BPSK prespreading symbols
2	1	1.8–14.4 Kbps, BPSK prespreading symbols
3	1	1.2–153.6 Kbps, QPSK prespreading symbols
4	1	1.2–307.2 Kbps, QPSK prespreading symbols
5	1	1.2–230.4 Kbps, QPSK prespreading symbols
6	3	1.2–307.2 Kbps, QPSK prespreading symbols
7	3	1.2–614.4 Kbps, QPSK prespreading symbols
8	3	1.2–460.8 Kbps, QPSK prespreading symbols
9	3	1.2–1,036.8 Kbps, QPSK prespreading symbols

process. The short code uses two sequences: an in-phase $PN_I(x)$ and a quadrature $PN_Q(x)$ sequence. For SR1 the codes are generated by the following LFSR polynomials:

$$PN_I(x) = x^{15} + x^{13} + x^9 + x^8 + x^7 + x^5 + x^0 \tag{8.3}$$

$$PN_Q(x) = x^{15} + x^{12} + x^{11} + x^{10} + x^6 + x^5 + x^4 + x^3 + x^0 \tag{8.4}$$

A maximal length y stage LFSR will have a period of $2^y - 1$. The output of the short code polynomials are slightly modified [12] by adding an extra zero to the sequence. The short code then has a period of 2^{15} or 32,768 chips, which is 80/3 ms or 27.667 ms. The SR1 long code is given by the following polynomial:

$$PN(x) = x^{42} + x^{35} + x^{33} + x^{31} + x^{27} + x^{26} + x^{25} +$$
$$x^{22} + x^{21} + x^{19} + x^{18} + x^{17} + x^{16} + x^{10} + x^7 + x^6 + \tag{8.5}$$
$$x^5 + x^3 + x^2 + x^1 + x^0$$

The long code has a period of $2^{42} - 1$ or 32,767 chips; this is approximately 41 days. At any given time a base station code clock is synchronized by loading the shift registers with the appropriate bits, equivalent to them being started at the GPS origin and then running through their respective sequences.

The SR3 I and Q PN sequences are based on the polynomial:

$$PN(x) = x^{20} + x^9 + x^5 + x^3 + x^0 \qquad (8.6)$$

and created by using different starting positions for I and Q and by truncating each sequence after 3×2^{15} chips.

For SR3 a 3.6864-Mcps-long code is created by multiplexing between three delayed copies of the SR1 code, each separated in time by $1/1.2288$ μsec.

8.2.3.3 Channel Structure

There are two basic classes of channels: common channels are shared among many users, and dedicated channels carry information for a single user. Figure 8.13 [11] illustrates the breakdown and structure of the forward channels for SR1 and SR3.

Common Channels

On the forward link CDMA2000 reuses several IS-95 common channels: pilot (F-PICH), paging (F-PCH), and sync (F-SYNC) (see Figure 8.14).

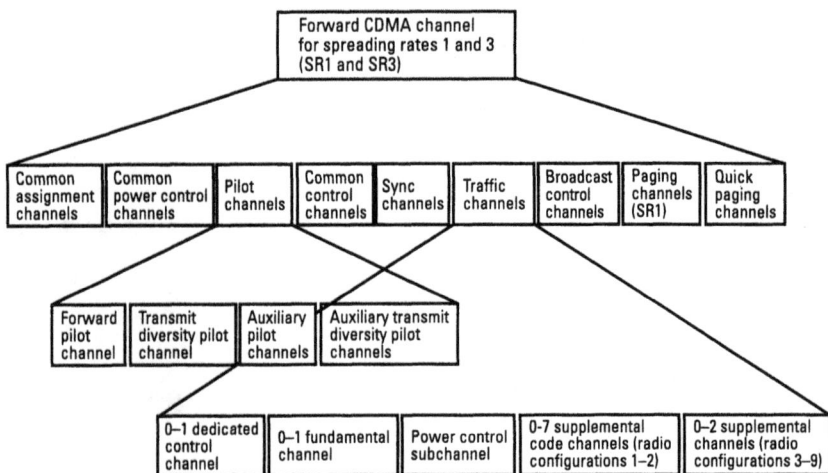

Figure 8.13   CDMA2000 channel structure architecture. (*Source:* 3GPP2, 2001. Reprinted with permission.)

Walsh codes are used for channelization on the forward link, where the pilot channel uses Walsh code 0, the sync channel is allocated to Walsh code 32, and paging channels can be assigned to channels 1 through 7.

Prior to establishing communication with a base station, the mobile searches for pilot channels to obtain initial synchronization; the pilot is then used for multipath channel estimation and coherent detection as well as frequency correction and, finally, for handover decisions. The pilot channel contains no information (i.e., all zero symbols) and must be receivable at all locations within the cell; it is, therefore, transmitted with more power (typically 15–25% of total) than other control channels.

Synchronizing to the pilot channel allows the mobile to decode the sync channel that contains a message specifying the phase (or PN timing offset) of the base station's short code; this information allows the mobile to despread a traffic channel with sufficient performance.

Paging channels are shared by all the mobiles in a cell; the channel is used to receive a range of information, including short message services, voice pages, system parameters, and other general cell broadcasts.

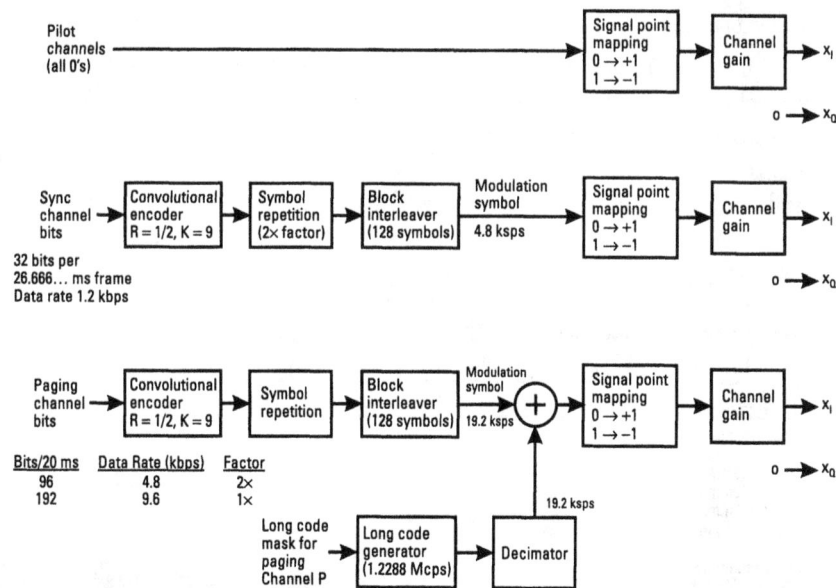

Figure 8.14 Pilot, sync, and paging channel structure for SR1. (*Source:* 3GPP2, 2001. Reprinted with permission.)

CDMA2000 adds several new forward common channels. Increased signaling capacity is provided by the broadcast control channel (F-BCCH) and common control channel (F-CCCH). The auxiliary pilot channel (F-CAPICH) is used for smart antenna applications and provides a phase reference for coherent demodulation of associated forward link channels. To help reduce mobile power drain the quick paging channel (F-QPCH) is an uncoded, spread, and on-off-keyed (OOK) modulated signal sent by the base station to wake up the mobile to receive a message on the paging channel. The final addition is a common power control channel (F-CPCH), which sends power control messages to multiple mobiles.

On the reverse link, channelization (separation of mobiles) is achieved by using long code masks (PN phase offsets). The IS-95 reverse access channel is reused by CDMA2000 for short signaling message exchanges, such as call origination, paging responses, and registration. The CDMA2000 reverse link adds a common control channel (R-CCH) and an enhanced access channel (R-EACH). These are used to increase the probability of a base station detecting a mobile by slotting bursts from different users and introducing coherence to provide receiver gain.

Dedicated Channels

The CDMA2000 forward dedicated channel set inherits the IS-95 fundamental channel (F-FCH) and the supplemental code channel (F-SCCH). The fundamental channel is variable rate; it includes the traffic channel (e.g., 13-Kbps QCELP voice) along with power control information. The supplemental code channel is part of RC1 and RC2 and operates in conjunction

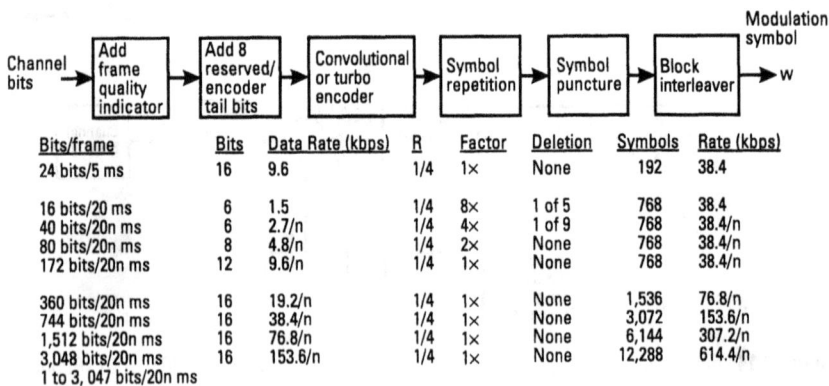

Bits/frame	Bits	Data Rate (kbps)	R	Factor	Deletion	Symbols	Rate (kbps)
24 bits/5 ms	16	9.6	1/4	1×	None	192	38.4
16 bits/20 ms	6	1.5	1/4	8×	1 of 5	768	38.4
40 bits/20n ms	6	2.7/n	1/4	4×	1 of 9	768	38.4/n
80 bits/20n ms	8	4.8/n	1/4	2×	None	768	38.4/n
172 bits/20n ms	12	9.6/n	1/4	1×	None	768	38.4/n
360 bits/20n ms	16	19.2/n	1/4	1×	None	1,536	76.8/n
744 bits/20n ms	16	38.4/n	1/4	1×	None	3,072	153.6/n
1,512 bits/20n ms	16	76.8/n	1/4	1×	None	6,144	307.2/n
3,048 bits/20n ms	16	153.6/n	1/4	1×	None	12,288	614.4/n
1 to 3, 047 bits/20n ms							

Figure 8.15   F-FCH and F-SCH structure for RC3. (*Source:* 3GPP2, 2001. Reprinted with permission.)

with the fundamental channel to provide higher data rates. The CDMA-2000 forward channel set is completed with the supplemental channel (F-SCH) and the dedicated common control channel (F-DCCH). The F-SCH and F-DCCH are portions of the fundamental channel for RC3 to RC9. The F-SCH is used in conjunction with a fundamental channel or a dedicated control channel to provide higher data rate services. The F-DCCH is used for the transmission of higher-level data, control information, and power control information. (See Figures 8.15 and 8.16.)

The reverse link includes reciprocal versions of the forward fundamental (R-FCH) and supplemental (R-SCH) channels. The CDMA2000 standard adds a reverse pilot channel (R-PICH) and a reverse common control channel (R-DCCH). The R-DCCH is reciprocal to the F-DCCH, and the R-PICH is functionally similar to the forward common pilot channel. The reverse pilot is an unmodulated direct sequence spread spectrum signal transmitted continuously by the mobile to provide the base station with a phase reference and allow it to coherently demodulate the mobile's transmission.

8.2.3.4 Spreading and Modulation

Previous figures illustrated the sequence of operations that is performed on channel bits to produce modulations symbols, W (Figure 8.15), followed by the conversion of modulations symbols to produce long code scrambled symbols, X (Figure 8.16), or quadrature symbols, X_I and X_Q (Figure 8.14). Figure 8.17 continues by illustrating the demultiplexing of long code scram-

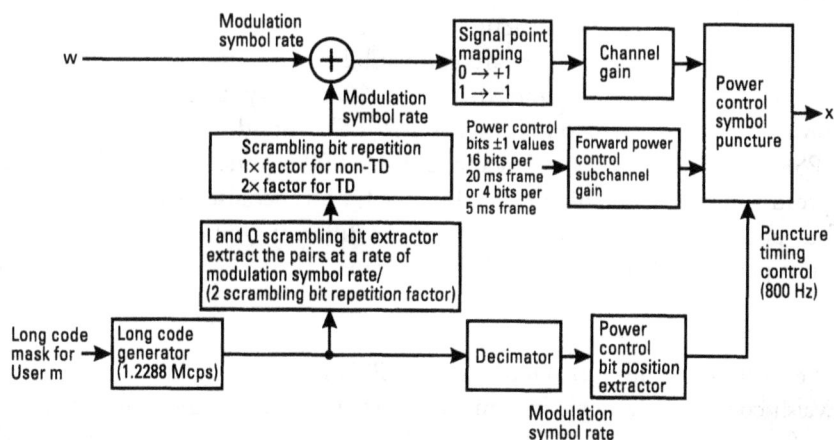

Figure 8.16 Long code scrambling, power control, and signal point mapping for forward traffic channels for RC3, RC4, and RC5. (*Source:* 3GPP2, 2001. Reprinted with permission.)

(a) Non-TD mode

(b) TD mode

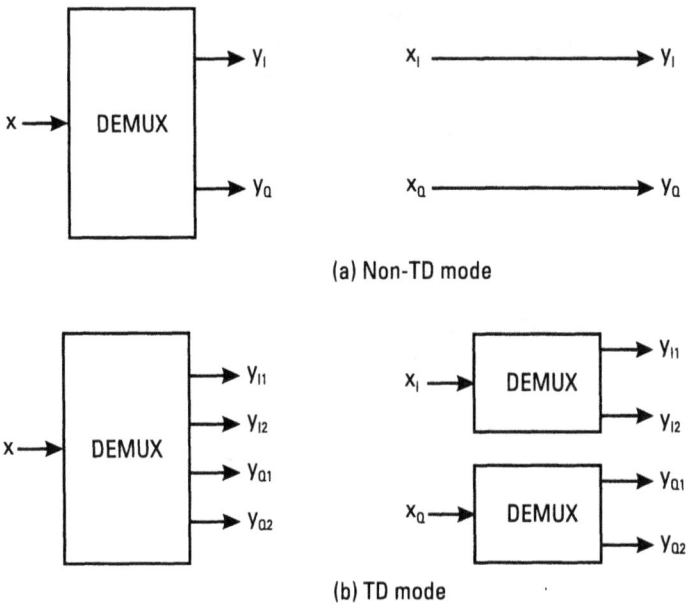

Figure 8.17 Demultiplexer structure for SR1. (*Source:* 3GPP2, 2001. Reprinted with permission.)

bled symbols X, X_I, and X_Q to produce in-phase (Y_I) and quadrature (Y_Q) symbols.

For SR1 the Y_I and Y_Q symbols are spread and modulated, as shown in Figure 8.18.

The demultiplexing structure, combined with the spreading and modulation design, provides backward compatibility. The result is that for IS-95 channels (Figure 8.14) X_Q and, therefore, Y_Q are always zero; in this case the spread and modulated on-air signal is QPSK spread superimposed on a BPSK code symbol stream. For the CDMA2000 channels (Figure 8.16) the spread and modulated on-air signal is true QPSK, because each point on the four-point QPSK constellation represents a different symbol; in other words, 2 bits per symbol are transmitted for this case.

8.2.3.5   Quasi-Orthogonal Codes

The IS-95A and IS-95B standards use BPSK prespreading and fixed 64-bit Walsh codes. The variable information rates supported by the different radio configurations of CDMA2000 require the use of variable-length codes. Walsh codes from 4 chips in length to 128 chips for SR1 and 256 chips for SR3 are

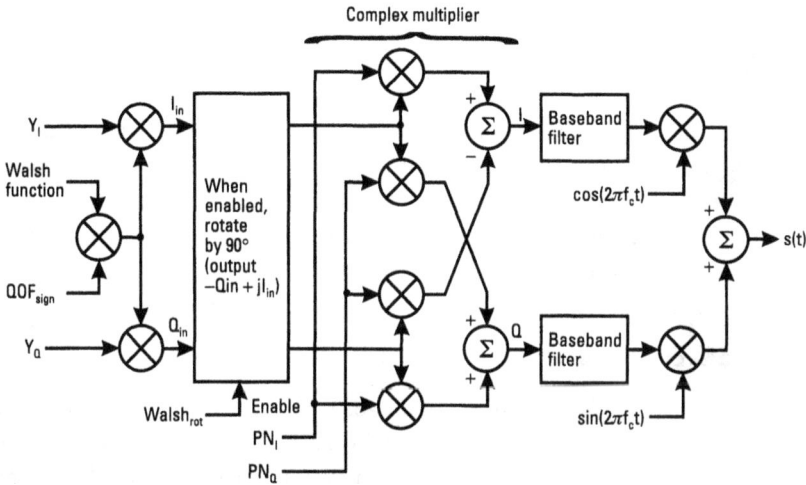

Walsh function = ±1 (mapping: '0' → +1, '1' → −1)
QOFsign = ±1 sign multiplier QOF mask (mapping: '0' → +1, '1' → −1)
Walsh$_{rot}$ = '0' or '1' 90°-rotation-enabled Walsh function
    Walsh$_{rot}$ = '0' means no rotation
    Walsh$_{rot}$ = '1' means rotation by 90°
The null QOF has QOF$_{sign}$ = +1 and Walsh$_{rot}$ = '0'.
PNI and PNQ = ±1 I-channel and Q-channel PN sequences
The null QOF is used for Radio Configurations 1 and 2

Figure 8.18 Forward link spreading and modulation for SR1. (*Source:* 3GPP2, 2001. Reprinted with permission.)

used. New codes (known as quasi-orthogonal codes) are created by using a nonzero sign multiplier quasi-orthogonal function (QOF) mask and a nonzero rotate enable Walsh function on the base Walsh codes (see Figure 8.19).

Function	Masking Function	
	Hexadecimal Representation of QOF$_{sign}$	Walsh$_{rot}$
0	00000000000000000000000000000000 00000000000000000000000000000000	W_0^{256}
1	7228d7724eebebb1eb4eblebd78d8d28 278282d81b41be1b411b1bbe7dd8277d	W_{130}^{256}
2	114b1e4444e14beeee4be144bbe1b4ee dd872d77882d78dd2287d277772d87dd	W_{173}^{256}
3	1724bd71b28118d48ebddb172b187eb2 e7d4b27ebd8ee82481b22be7dbe871bd	W_{47}^{256}

Figure 8.19 Masking functions for SR1 and SR3 MC mode. (*Source:* 3GPP2, 2001. Reprinted with permission.)

8.2.3.6 Future CDMA2000 Modes

The CDMA2000 standard is continuing to evolve, with extensions that offer increased spectrum efficiency and potentially higher data rates. The Evolution Data Only (EVDO) standard is being supported by Qualcomm; as the name suggests, this offers increased rates for packet type data only services. The Evolution Data and Voice (EVDV) standard is being developed by several companies, including Texas Instruments, Nokia, and Ericsson; it supports both data- and packet-based voice services.

8.2.4 GSM Evolution

As a 2G standard GSM supports basic voice services but can also provide user access to various data services up to 9.6 Kbps. GSM can be operated in many bands, with the 900-MHz allocation of 890–915 MHz for the uplink and 935–960 MHz for the downlink being widespread throughout the world. The GSM air interface uses a mix of frequency division multiple access (FDMA) and time division multiple access (TDMA). Each 25 MHz of transmit and receive spectrum is divided into 125 carriers of 200-kHz bandwidth. Each cell in the network is assigned a number of these carriers, and the basic cellular principle of carrier reuse in nearby cells is applied. A user can be frequency hopped between the carriers in a cell; this technique can improve system performance by averaging down frequency selective fading and cochannel interference.

Each carrier is shared in time, where the basic unit is a time slot of 0.577 ms in length. Figure 8.20 [7] illustrates the hierarchical structure, starting with a basic TDMA frame (eight time slots) that is built up into 26-frame or 51-frame multiframes, followed by 1,326 frame superframes, and finishing with 2,715,648-frame hyper-frames. Each time slot contains 156.25 bits, which equates to a total carrier capacity of 270.8 Kbps.

A physical channel is, therefore, defined as a sequence of TDMA frames, a time slot number (modulo 8), and a frequency hopping sequence.

Bits are modulated onto the carrier using GMSK; this modulation scheme allows for 1 bit per symbol. At even-numbered symbols the polarity of the I channel conveys the transmitted data, while at odd-numbered symbols the polarity of the Q channel conveys the data. The orthogonality between I and Q simplifies detection algorithms and, hence, reduces the power consumption in the receiver.

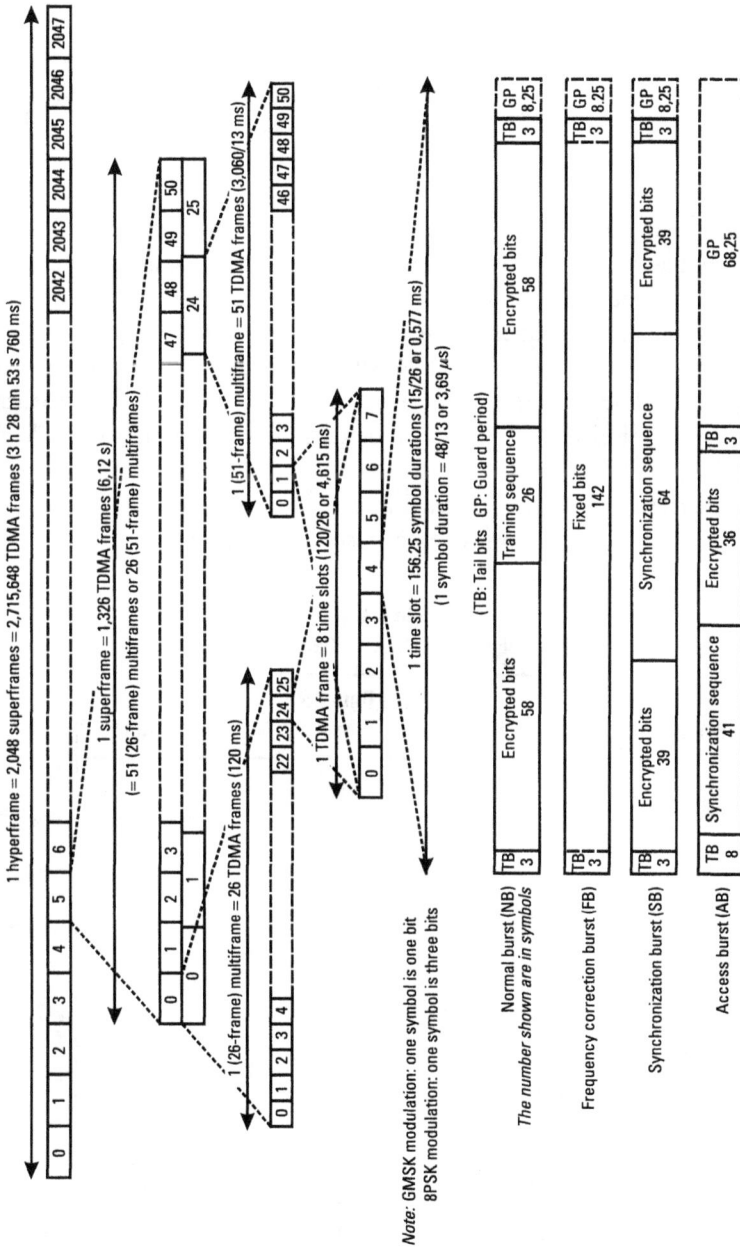

Figure 8.20 GSM time frames, time slots, and bursts. (©ETSI 2001. Further use, modification, or redistribution is strictly prohibited; ETSI standards are available from http://pda.etsi.org/pda/ and http://www.etsi.org/eds/. Reprinted with permission.)

The 3GPP has defined an evolutionary path for the upgrade of the second-generation GSM digital standard. This evolution provides higher data speeds than could be previously achieved, with the ability for always-on type data connections. The first stage is an upgrade to the general packet radio service (GPRS), with even greater speeds being possible during a subsequent upgrade to the enhanced data for GSM evolution (EDGE) standard.

As an extension of GSM, GPRS uses the same GMSK modulation approach but adds several more complex coding schemes to the channel coding part of the system. As shown in Table 8.3, CS-1 to CS-4 provide varying data rates when a user is allocated use of a single time slot.

Additional bandwidth is possible with GPRS by allocating more than one time slot. More time slots can be allocated in the downlink versus the uplink (asymmetrical allocation), which better suits applications such as Web browsing.

To increase data rates further, EDGE takes the step of increasing the complexity of the modulation scheme from GMSK to 8 PSK. With eight

Table 8.3
GPRS and EDGE Data Rates

Evolution	Coding Scheme	Data Rate (Kbps)	Modulation
GPRS	CS-1	9.05	GMSK
GPRS	CS-2	13.4	GMSK
GPRS	CS-3	15.6	GMSK
GPRS	CS-4	21.4	GMSK
EDGE	MCS-1	8.8	GMSK
EDGE	MCS-2	11.2	GMSK
EDGE	MCS-3	14.8	GMSK
EDGE	MCS-4	17.6	GMSK
EDGE	MCS-5	22.4	8 PSK
EDGE	MCS-6	29.6	8 PSK
EDGE	MCS-7	44.8	8 PSK
EDGE	MCS-8	54.4	8 PSK
EDGE	MCS-9	59.2	8 PSK

points in the constellation (8 points = $2^{3\text{ bits}}$), spectrum efficiency is increased by a factor of three. The theoretical maximum throughput for all eight time slots of a 200-kHz GSM carrier using the MCS-9 coding scheme then becomes 8 × 59.2 Kbps, or 473.6 Kbps; actual payload will only be approximately 384 Kbit/sec after subtracting all overhead (error correction and so on) bits.

Although EDGE does not reach the 3G rate of 2 Mbps, at 384 Kbps it will be able to support substantial data type services and compares well with WCDMA in terms of spectrum utilization.

8.3   Software Defined Radio Examples

This section starts by covering two examples of software radio products that are available as platforms, followed by two examples where complete applications are incorporated into black box base stations. The section concludes with a review of two research, based software radio projects.

8.3.1   Frameworks and Platforms

This chapter defines a software radio framework or platform as the combination of hardware and software that broadly meets the DOCSRA or JTRS specifications and is suitable for software radio applications development.

8.3.1.1   SpectruCell

Advanced Communications Technologies commenced operations in 1999 to develop the SpectruCell software defined radio product. Initially SpectruCell was intended to be delivered to the market as a black box BTS (with a 2G, 2.5G, or 3G application) product. However, after refining the company's plans, SpectruCell is being made available as a software defined radio platform framework for use by application developers.

SpectruCell is primarily targeted to 3G base station applications but can also be applied to other applications in the defense and communications industry. Application software has been developed for GSM, IS-95, and CDMA2000 as part of a program to validate the framework design. Application developers have the task of using the framework to design and build the application software. They must then integrate and test the combined framework and application, including any requirements for type approval as required by the country where the equipment will be deployed.

The aim of the SpectruCell framework is to provide application developers with a platform of hardware and middleware capable of implementing current and future mobile cellular base station protocols. Figure 8.21 illustrates the layered approach, with hardware at the base of the platform playing host to the SpectruCell Software Radio Operating System (SSROS). The SSROS includes COTS operating systems and SpectruCell core middleware. The example CDMA2000 application shows various software resources interfacing with the middleware via a number of SpectruCell Application Programming Interfaces (SCAPIs).

Figure 8.21 also shows a functional view of the software defined radio by indicating partitioning through each of the application, middleware, and hardware layers. For example, IF software interfaces with the middleware, which, in turn, interfaces with the IF hardware.

The hardware platform is available in two form factors: one is suitable for implementing low-capacity base stations (e.g., micro- and picocellular), and the other is tailored for much higher capacity situations (e.g., macro cellular).

SpectruCell Framework

The key components of the framework are the hardware platform, the SpectruCell Software Radio Operating System, and the system RTOSs.

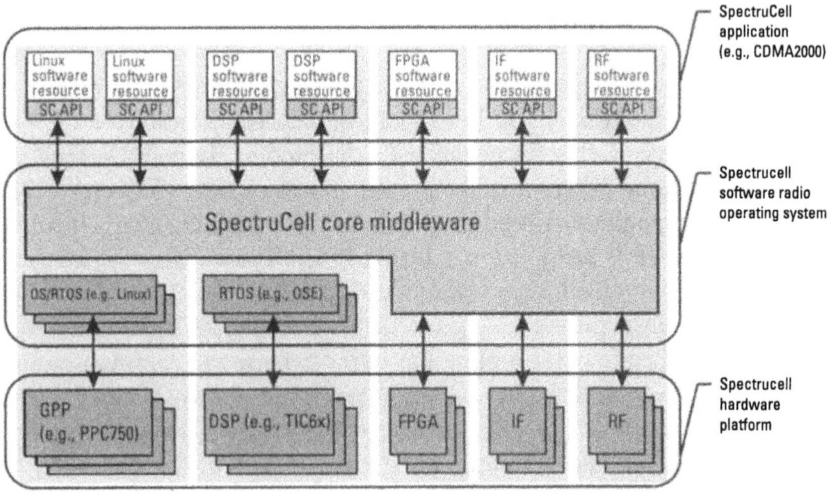

Figure 8.21 SpectruCell framework and example application. (*Source:* ACT Australia, 2001. Reprinted with permission.)

SpectruCell Hardware. The macro- and microplatforms have been designed to efficiently implement cellular mobile BTS applications across the range of 2G and 3G air interfaces (e.g., GSM 900, GSM 1800, IS-95, CDMA2000, and UMTS). The macro form-factor. "MacroSpec," is designed to be efficiently deployed at BTS sites where a large cell radius is required or extra signal processing power is needed to perform capacity enhancing algorithms—smart antenna or multiuser detection (MUD). The micro form-factor, "MicroSpec," is designed to be efficiently deployed in smaller radius, lower-capacity cell sites.

The form-factors share common hardware processing components (ADCs, DACs, digital frequency upconverters, digital frequency downconverters, FPGAs, DSPs and reconfigurable processors) to maximize the transparency of software development and software module reuse.

The form-factors also share common wideband multicarrier analog front ends (analog upconverter, analog downconverter, low noise receive amplifier, and multicarrier transmit power amplifiers).

MacroSpec The MacroSpec platform architecture is designed to cost effectively meet the requirements of macro-sized cells. Cell size as specified by the ITU [13], is defined in Table 8.4.

Since macrocells have cell radiuses in the order of kilometers, they tend to be deployed during the early stages of a new network to quickly provide coverage to a large area (approximately 3–1,000 km^2) with a minimal number of base stations. Larger link distance implies bigger power amplifiers and higher antenna systems. As a result, macrocells need to provide a high level of reliability, especially during the start of a network deployment, since they tend to be a single point of failure in the system. As the density of users increases, these macrocells start to become overloaded, and network capacity is usually improved by adding micro- and picocells in hot spots (localized

Table 8.4
Cell Size Classification

Cell Classification	Cell Radius
Pico	$r < 100m$
Micro	$100m < r < 1,000m$
Macro	$1,000m < r < 35\ km$

areas of high density). Typical microcell locations include central business district street corners, shopping center parking garages, and in-building situations. The macrocells are left to provide umbrella coverage and also serve to improve reliability by providing fall-back capacity when a microcell fails or is required to be taken off-line for a period of maintenance or upgrade.

The hardware architecture is optimized for the combination of user capacity and larger cell radius and provides modular hardware components interconnected by flexible switching fabrics. Expandability and redundancy are key features, and the system incorporates full hot-swap and N + 1 redundancy throughout.

The MacroSpec architecture shown in Figure 8.22 illustrates the three key components: hardware platform, operating system, and application software. The hardware is partitioned into three functional blocks: analog RF, digital intermediate frequency (IF), and digital signal processing (SP). IF, SP, system controller, and the network interfaces mount in a standard 19" rackmount Compact PCI (cPCI) chassis.

Each RF block represents a complete wideband RF receive or transmit chain capable of processing up to 20 MHz of spectrum. The RF chains are responsible for A/D conversion, D/A conversion, analog up- and downcon-

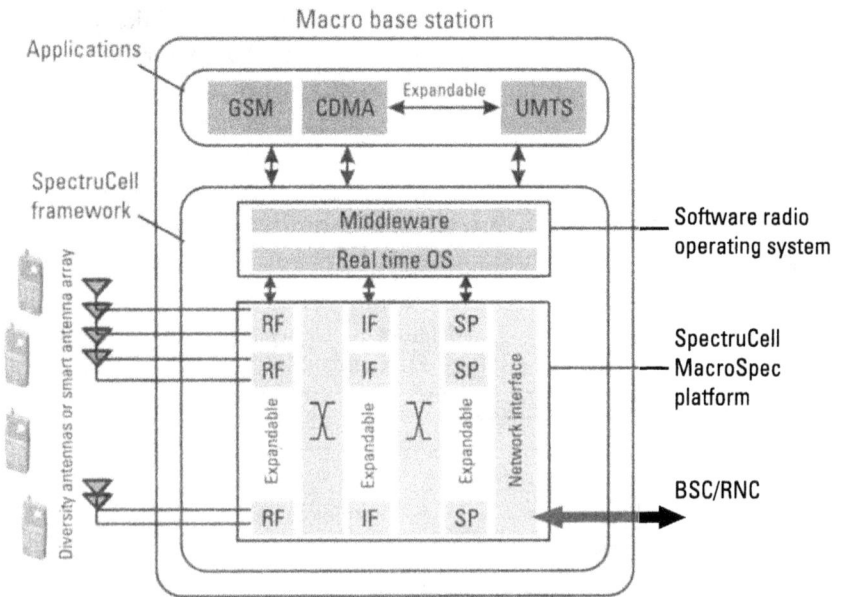

Figure 8.22 Macroarchitecture. (*Source:* ACT Australia, 2001. Reprinted with permission.)

version to a common digital IF, low noise receive amplification, and power amplification. The RF chains are connected by a flexible switching matrix to the digital IF section. This arrangement allows the digital IF and SP hardware to process signals from a flexible range of sources, including spatially diverse antennas or from different sectors.

Each IF block represents a high-capacity IF processing card, which is capable of operating on multiple instances of receive and transmit RF carriers. On the receive side for a given instance, IF processing includes selection of a carrier at the common digital IF frequency as well as filtering, decimation, and frequency downconversion to baseband. On the transmit side, IF processing includes digital frequency upconversion from baseband, rate matching, and carrier combination to present a single wideband multicarrier source to a D/A converter.

The SP blocks in Figure 8.22 represent high-capacity generic digital signal processing cards. The SP cards are connected via a flexible switching system to the IF cards so that any baseband digital data stream from the array of IF cards can be switched to any of the SP cards. The SP cards are the prime layer one resource in the BTS, being responsible for sourcing traffic (data and voice) and control channels from the BSC and encoding and modulating them as per the resident layer one air interface applications (e.g., CDMA2000, UMTS, and so on).

Figure 8.22 illustrates the separation (abstraction) of the application software from the hardware via a layer of middleware.

MicroSpec   The MicroSpec hardware platform, illustrated in Figure 8.23, is a more integrated version of the MacroSpec hardware; it is designed for high user density and small cell area traffic scenarios requiring lower-power RF PAs. The SDPP card includes IF, SP, system controller, and network interface functionality. This approach is designed to meet the requirements of lower-cost BTSs, which occupy less space and consume less power.

The SDPP cards are standalone (not rack mounted), and more capacity can be added with additional SDPP cards; extra RF hardware is usually not necessary.

The MicroSpec is as flexible from a software point of view as the MacroSpec system and can host multiple air interface applications and other capacity enhancing applications such as smart antennas.

Software Radio Operating System   The SpectruCell Software Radio Operating System builds on the technical features of the SpectruCell hardware platform and is aimed at significantly reducing the engineering effort and time-to-

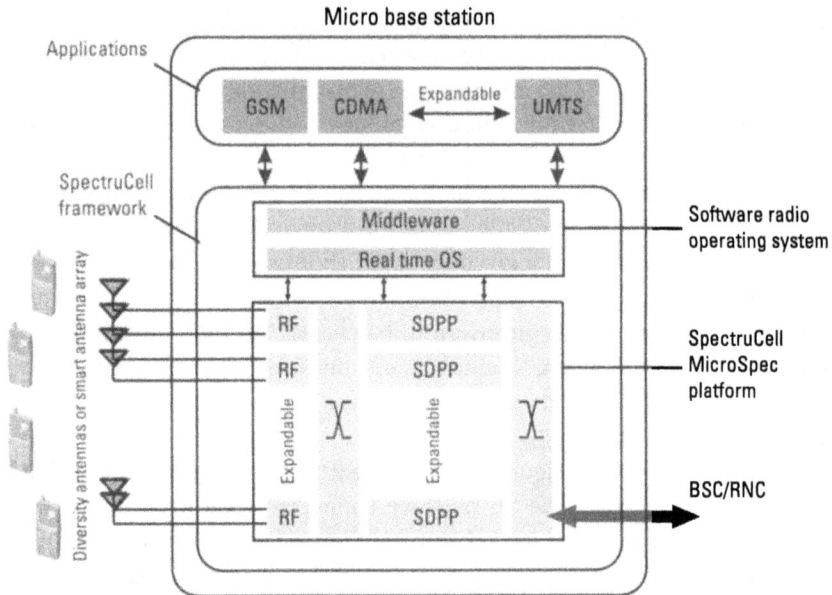

Figure 8.23 Microarchitecture. (*Source:* ACT Australia, 2001. Reprinted with permission.)

build applications. Application developers need only concentrate on the important application layer code without being concerned with the complexities inherent in developing distributed real-time systems.

The SpectruCell Software Radio Operating System is made up of the following key components:

- Core middleware;
- Middleware API;
- Real-time operating systems.

Several development tools are provided for application developers, including SC-IDE, SC-Probe, SC-ApplicationBuilder, SC-MessageBuilder, and SC-SimulationSuite. Supported third-party development tools include the GNU integrated compiler/debugger tool suite, Code profiler, Telelogic SDL/TTCN tool suite, and Texas Instruments' Code Composer.

Core Middleware   The Core Middleware architecture, depicted in Figure 8.24, is based on the SDRF Distributed Object Computing Software Radio Architecture [14] and has several key enhancements for performance and stability.

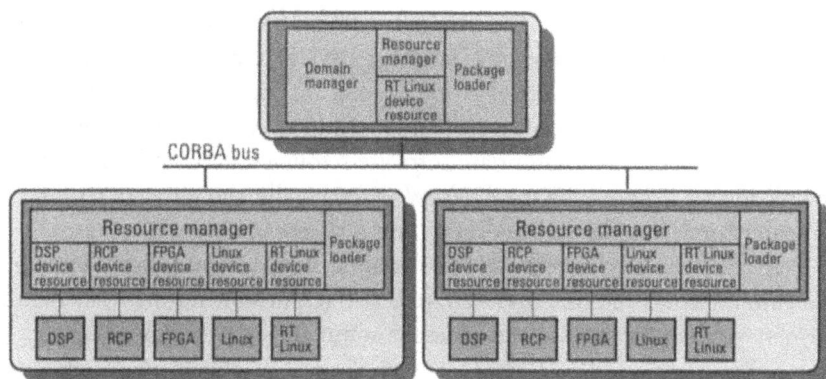

Figure 8.24 Core Middleware architecture. (*Source:* ACT Australia, 2001. Reprinted with permission.)

The Core Middleware *DomainManager* is responsible for coordinating distributed application loading and management. It can manage multiple applications on the same hardware simultaneously and manages resource allocation and other application services such as virtual circuit and error management.

The *ResourceManager* is responsible for managing local hardware devices. It creates a *DeviceResource* for each physical hardware device and registers its capabilities with the *DomainManager*. Each *DeviceResource* loads and manages application software on the device.

The *PackageLoader* is responsible for locating the appropriate application software. Each *PackageLoader* communicates with its parent to locate and download the appropriate type and version of application software. It also performs code validation and verification using checksums and digital certificates.

An application is made up of a number of distributed software resources. There are different types of software resources for different devices, for example:

- LINUX software resource (single process running under LINUX);
- Real-time LINUX software resource (thread running under RT LINUX);
- MQX software resource (task running under MQX on a DSP);
- FPGA software resource (complete configuration for a particular FPGA).

The application developer creates a configuration file using SC-Application Builder, which specifies the hardware devices required by the application and the software resources to be loaded on each device. Devices are allocated to the application at run time by the *DomainManager* and the *ResourceManagers*. The appropriate software resources are then located (and downloaded if required) by the *PackageLoader*. Finally, the software resources are loaded onto the physical devices by the *DeviceResource*.

In addition to controlling resource allocation, locating, downloading, and running code, the core middleware is also responsible for life-cycle management of the application. Management commands are sent to the application components via the *DomainManager*. Typical management commands include run, halt, enable, disable, and configure.

The central resource allocation and management capabilities of the design allow multiple applications to share one hardware platform. Modular construction allows developers to choose the framework features needed for the target processor. The core framework allows fully managed remote download and distribution of software resources with support for TMN and SNMP style network management interfaces.

Middleware API   The SpectruCell middleware application programming interface (API) is an extensive set of application independent libraries that work closely with the core middleware components to ease application development and provide the required run-time performance.

The API is constructed of virtual circuit communications, state machine framework, error management, configuration, life cycle, concurrency interface, message encoding/decoding, and utilities libraries.

Virtual Circuit Communications Library   The virtual circuit communications part of the API provides a logical point-to-point data connection between local and/or remote software resources. It is designed to provide high-performance intersoftware resource communication from any device to any other device within the platform. The application software is abstracted from the underlying hardware and operating system and the mechanics of how the data is communicated. This mechanism provides as small an overhead as possible compared with applications using native platform facilities. (See Figure 8.25.)

The library facilitates fail-over and redundancy between dynamically creatable application software resources. It also allows probing of intersoftware resource communications for debugging and analysis purposes.

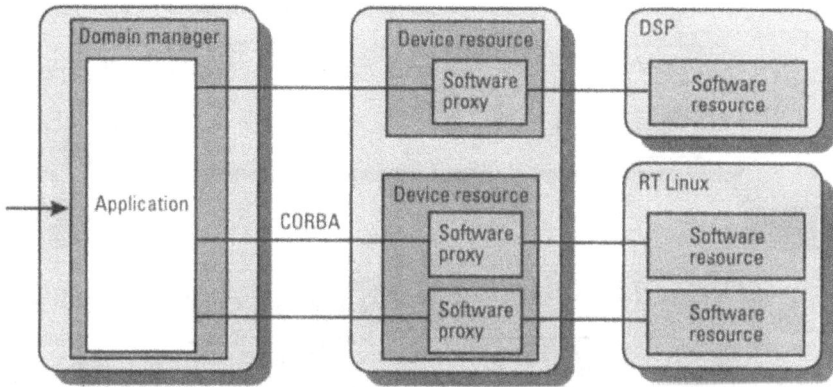

Figure 8.25 Virtual circuit architecture. (*Source:* ACT Australia, 2001. Reprinted with permission.)

State Machine Framework Library The state machine framework library provides a set of abstract base classes designed to help the application developer create complex high-performance state machines. Typical applications include layer two and layer three protocol stacks, control algorithms, and so on.

Services provided by the library include high- and low-resolution timers, message encoding and decoding, parallel state machines, automatic creation and deletion of state machine instances, and memory pool management.

Error Management Library The error management library provides a convenient and simple mechanism to manage and report alarms and errors. Alarms can be raised within the application software resources and will propagate through the platform to the network management interface if desired. The majority of the error fields required to generate TMN style alarms can be filled in automatically by the error management library.

Configuration, Life-Cycle, and Concurrency Interface Library This library provides the main interfaces to the core middleware. It consists of a set of abstract base classes used by the application developer to implement active software resources.

Each active software resource becomes a task, thread, or process (depending on the target operating system) and has common entry points and skeleton functions for implementing the enable, run, halt, disable, and configure actions.

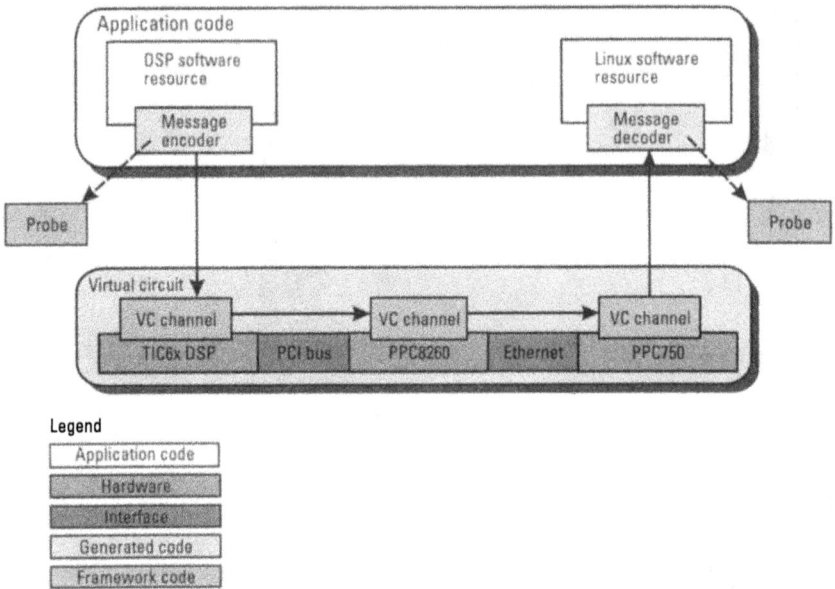

Figure 8.26 Message routing. (*Source:* ACT Australia, 2001. Reprinted with permission.)

Message Encoding/Decoding Library The message encoding/decoding library consists of a compile time code generator and a set of functions for creating, encoding, and decoding messages. (See Figure 8.26.)

Highly optimized C or C++ code can be generated for very complex bit-packed messages. Test code for each message is also generated, and the applications developer can supply arbitrary test vectors to the test suite, allowing off-line testing of the generated code. This allows application developers to describe complex protocol messages in a simple XML-based language with support for optional fields, non-byte-aligned fields, and arbitrary-length fields.

The generated code integrates seamlessly with the state machine framework library and the core active software resources. The core middleware uses the same system for all of its internal messages, allowing integrated probing and debugging of both application-specific messages and internal messages. (See Figure 8.27.)

Real-Time Operating Systems SpectruCell middleware supports a range of target processors, including PowerPC-based CPUs, Intel-compatible CPUs, and TI DSPs. Application development can be undertaken using the

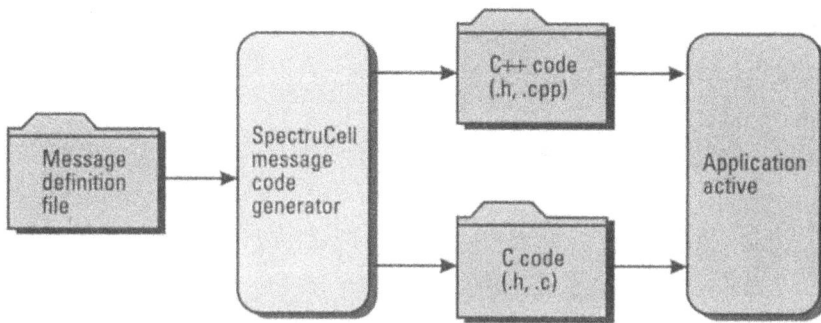

Figure 8.27 Message generation. (*Source:* ACT Australia, 2001. Reprinted with permission.)

Windows, LINUX, Solaris, or HP-UX host environments. Target operating systems for CPU targets include LINUX, RT LINUX, and VxWorks; for DSP targets MQX and OSE are supported.

Base Station Development Kit Application developers are supported by a base station development kit (BDK); the kit is a set of tools and development methods that define the software development environment. The tools are glued together by a fully integrated development environment (IDE); Figure 8.28 depicts a view of the IDE.

Application compilation is performed in the native format of the target device to ensure a minimal run-time application overhead. The compiler and debugger environment is based on the GNU tool suite; it allows the user to choose the host OS and the target to be debugged remotely. Debugging is also enhanced by the provision of a run-time data capture facility built into the core framework. This allows application developers to capture and view all messages exchanged between software components in real time.

The tool set includes algorithmic and hardware-level simulators to ensure that software is validated in the user environment prior to down load to the target hardware.

Test Mobile Emulator Developing a new base station software application with a minimum of effort requires completely functional emulators for each of the BTS's major interfaces—that is, the network side (BTS/RNC) and the air interface side (mobile/cell phone). Network emulator side and air interface side test equipment is readily available in the industry; however, there are few air interface emulators suitable for debugging and testing during the development phase of a BTS.

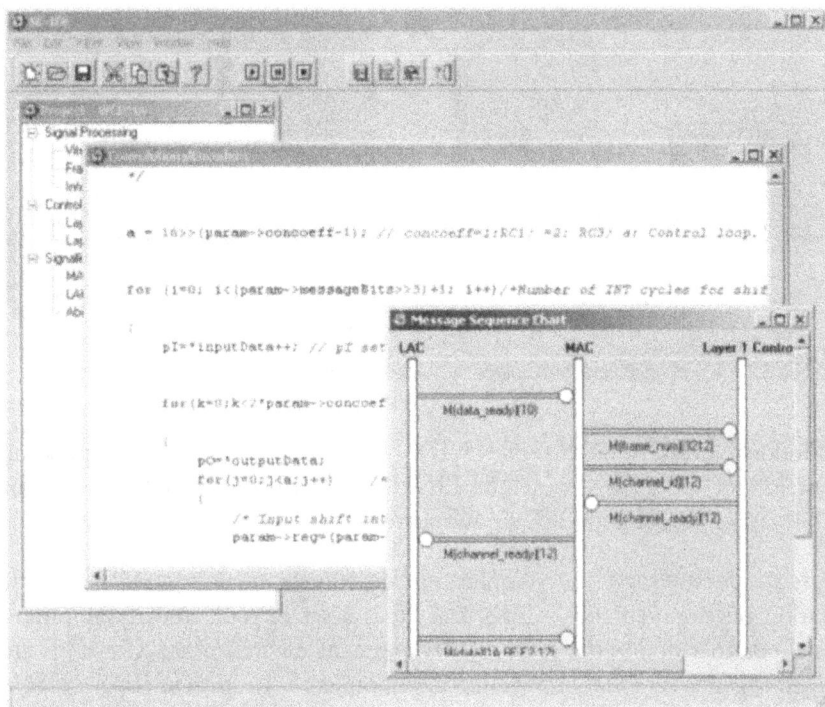

Figure 8.28 Integrated development environment. (*Source:* ACT Australia, 2001. Reprinted with permission.)

The BDK solves this problem by providing a software radio-based test mobile emulator (TME); see Figure 8.29.

The TME completely emulates the functionality of a mobile phone and provides hooks into each layer of the protocol stack. The capture and debug software can be set to either passively gather message data or to intervene in the normal flow of the state machine and bypass states or inject errors. Using the emulator during the development of application software for the base station reduces the need to step the base station through each stage of the life cycle to test any new functionality. This increases the degree of parallel development by allowing the early testing of functions that are normally used much later in the cycle of the base station state machine.

Application-Specific Performance

The SpectruCell framework is designed to host 2G and 3G mobile cellular air interfaces and will also be able to support changes to these specifications

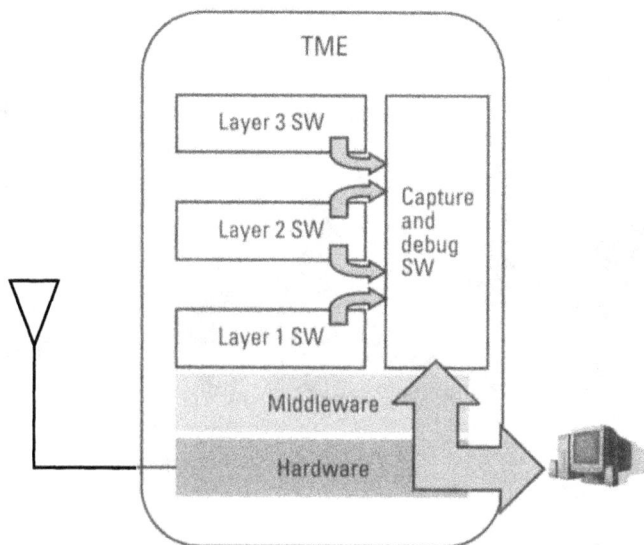

Figure 8.29 Test mobile emulator. (*Source:* ACT Australia, 2001. Reprinted with permission.)

and eventual upgrade to 4G when it becomes available. Figure 8.30 illustrates a commercial IS-95B CDMA mobile phone ready to establish or receive a call after registering with the SpectruCell software defined BTS.

Table 8.5 details the application-specific parameters for a range of air interfaces, considering that the framework is hosting BTS software radio applications. The "SDR front-end BW" parameter is the sampled digital bandwidth containing multiple RF channels. The "minimum number of users" figures assumes 9.6 Kbps users for a minimum system configuration (one IF and one SP card with a minimum set of hardware). The "maximum number of users" figures assumes 9.6 Kbps users for a maximum system configuration (a full cPCI shelf).

8.3.1.2   AdapDev SDR

Mercury Computer started in 1983 developing signal processing hardware primarily for defense applications; since that time they have branched out into medical imaging and telecommunications systems. They offer components based on industry standards, such as VME, cPCI, Race++, and RapidIO. The company has recently developed a software radio platform, AdapDev SDR, in accordance with the JTRS Specification [15]. The platform's hardware architecture is illustrated in Figure 8.31.

Figure 8.30 SDR BTS demonstration. (*Source:* ACT Australia, 2001. Reprinted with permission.)

The hardware architecture uses an industrial PC chassis to house four PCI segments and can be configured to include two ADC boards, one DAC board, and spare slots for network or serial I/O in segment one and additional processing modules in segment three. All cards interconnect to a host CPU via PCI, and the processing modules (VantageRT 7400) are also interconnected by RACE++. A summary of the system's specification is provided in Table 8.6.

On the software side AdapDev SDR provides a suite of software components for applications development, including run-time environment software, development tools, board drivers, middleware, and application examples in the C language.

The platform operates with the Mercury real-time MC/OS operating system. The OS has a nanokernal optimized to support high-bandwidth, low-latency multiprocessor operations. Middleware follows the JTRS recommendation and uses CORBA; the product is provided with the ORBExpress ORB.

Table 8.5
Macrospec Application-Specific Specification

	GSM 900	GSM 1800/1900	IS-95	CDMA 2000	UMTS
Air Interface >					
Physical					
Size Per Rack (m)			$1.8 \times 0.6 \times 0.6$		
Weight			Configuration Dependent		
Power					
Nominal Supply Voltages			220/110 VAC 50–60 Hz/–48 VDC		
Typical Power Consumption per User	50W	50W	20W	20W	20W
General					
Frequency Range (MHz)			As per standard		
SDR Front End BW	5 MHz	20 MHz	20 MHz	20 MHz	20 MHz
Omni Cell Capability	Yes	Yes	Yes	Yes	Yes
Sectored Cell Capability	Yes	Yes	Yes	Yes	Yes
Typical Sectored Configuration	3S/16C	3S/16C	3S/4C	3S/4C	3S/4C
Typical Capacity	0–100 E/km^2	0–100 E/km^2	0–100 E/km^2	0–100 E/km^2	0–100 E/km^2
Minimum Number of Users	128	128	128	128	32
Maximum Number of Users	512	512	512	512	128
Redundancy	Full N + 1				
Hot-Swap	Yes cPCI H110				
Downlink					
PA Multicarrier Capable	Yes	Yes	Yes	Yes	Yes

Table 8.5 *(continued)*
MacroSpec Application-Specific Specification

	Air Interface >				
	GSM 900	**GSM 1800/1900**	**IS-95**	**CDMA 2000**	**UMTS**
Downlink (cont'd.)					
PA Typical Carrier Power	20/40W	40/80W	40/80W	40/80W	40/80W
PA Maximum Carrier Power	80W	80W	80W	80W	80W
TX Diversity	N/A	N/A	Optional	Optional	Optional
Uplink					
Receive Sensitivity			As per standard		
Dynamic Range			As per standard		
SFDR			As per standard		
Rx Diversity	Yes	Yes	Yes	Yes	Yes
Interfaces					
Control	GSM-Compliant Abis	GSM-Compliant Abis	CDMA2000 Abis	CDMA2000 Abis	UMTS Iub
Console	Serial (DCE RS232)				
BSC/RNC Physical	Ethernet 10/100BaseT				
BSC/RNC Physical	8X T1 100 Ohm or 8X E1 75/120 Ohm				
BSC/RNC Physical	ATM (IMA over E1/T1) or STM-1 155 Mps, AAL2, AAL5 (IP support)				

Run-time environment software includes a scientific algorithm library with 400 routines implementing common digital signal processing mathematical operations. PAS is a library of functions for managing concurrent processes and distributed computation; this performs a key middleware function of hiding low-level data movement details from the application developer. A trace analysis tool and library are also provided to aid system-level debugging and analysis.

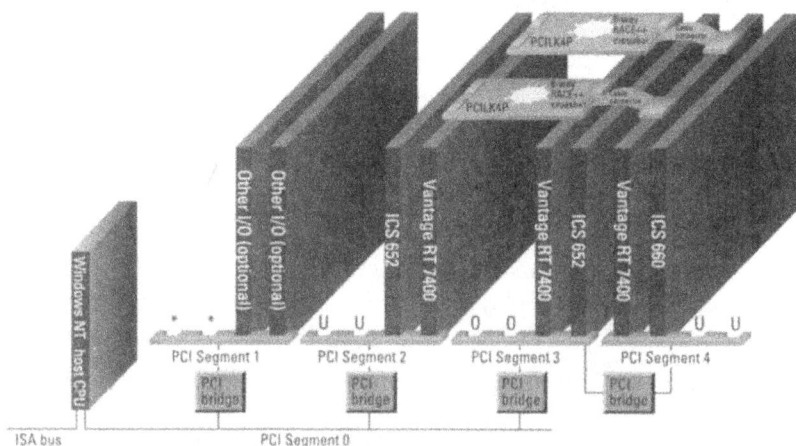

* = Open for additional network cards or serial I/O.
U = Unusable. These slots must remain empty.
O = Open for optional Vantage RT 7400 modules.

Figure 8.31 AdapDev SDR hardware architecture (*Source:* Mercury Computer, 2002. Reprinted with permission.)

Table 8.6
AdapDev SDR System Specification

Parameter	Specification
Beam Forming	Up to four element beam-forming transmit/receive array or dual-channel diversity antenna with analog I and Q
IF	Maximum IF frequency of 32 MHz
IF Bandwidth	IF bandwidth from 100 Hz to 5 MHz, maximum analog channel bandwidth of 5 MHz with 3 times oversampling
Number of Channels	Between 1 and 16 digital I and Q baseband channels, depending upon downconverter selection
ADC Board	Two analog input channels, 14-bit converter, 65-MHz sampling frequency, SNR 72 dB, SFDR 90 dB
DAC Board	Four analog output channels, 14-bit converter, 65-MHz sampling frequency, SNR 70 dB, SFDR 80 dB

8.3.2   Base Transceiver Stations

SpectruCell and AdapDev SDR are platforms designed to be used by application developers to create the final application software, integrate it with the platform, and produce a finished product. We will now cover two examples of complete black box software radio products: a 3G and a 2G base station.

8.3.2.1   Flexent OneBTS

Lucent Technologies is a major mobile cellular infrastructure manufacturer with more than 50,000 deployments of its IS-95 CDMA base station. The company has developed the Flexent OneBTS [16] as a UMTS/WCDMA Node B. The layout of hardware components within the base station is depicted in Figure 8.32.

The cabinet (or rack) for an indoor configuration measures 1.88m high by 0.6m wide by 0.6m deep and can be scaled to support anywhere between single carrier omnidirectional configurations to multicarrier multisectored ones. One cabinet can be expanded to a maximum of six 40W UMTS radio carriers spread across three sectors; these carriers can also be configured for transmit diversity by reducing each to 20W. The base station uses an IF bandwidth of 5 MHz, and baseband processing is performed by a combination of discrete DSPs and ASICs; in the future the DSP function may be integrated into the ASIC.

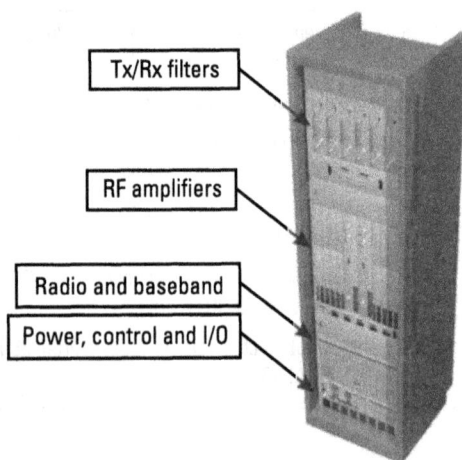

Figure 8.32 Flexent OneBTS Node B architecture. (*Source:* Lucent, 2002. Reprinted with permission.)

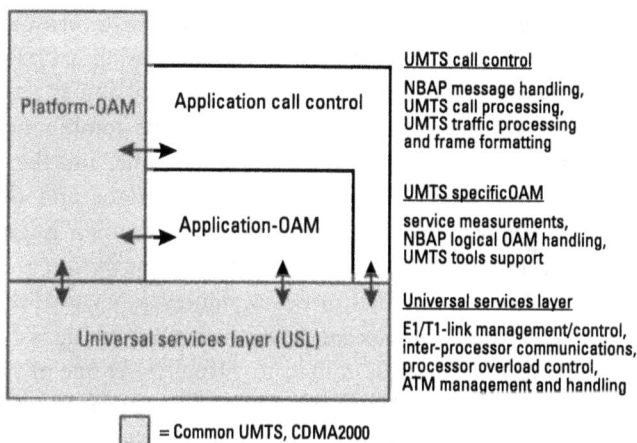

Platform OAM

Significant set of OAM Capabilities reused from Lucent CDMA products: system initilization, diagnostics, program storage and upgrade, recovery, reconfiguration, etc.

Platform-OAM

Application call control

Application-OAM

Universal services layer (USL)

UMTS call control

NBAP message handling, UMTS call processing, UMTS traffic processing and frame formatting

UMTS specificOAM

service measurements, NBAP logical OAM handling, UMTS tools support

Universal services layer

E1/T1-link management/control, inter-processor communications, processor overload control, ATM management and handling

= Common UMTS, CDMA2000

Figure 8.33 Flexent OneBTS Node B software architecture. (*Source:* Lucent, 2002. Reprinted with permission.)

A block diagram of the OneBTS call control and operations adminis-tration and maintenance (OAM) software architecture is shown in Figure 8.33. The platform OAM and universal services layer (USL) is common to UMTS and CDMA2000. Communications between the Node B's processor cards and baseband cards are transacted via the USL that runs on each card.

Pressure is mounting in countries with very dense 2G networks to min-imize new base station sites (particularly towers), and the idea of sharing air interface infrastructure is gaining momentum. In response to this the OneBTS is designed to accommodate multiple frequency allocations, which can be logically segregated through software.

8.3.2.2 AdaptaCell

Airnet is headquartered in Florida and was the first company to develop a software defined BTS for the GSM market in 1997. The product range includes the AdaptaCell BTS and the AirSite back-haul free BTS. An Adap-taCell can be connected back to a BSC via a single E1/T1 line, and the AdaptaCell can then used with up to 12 other AirSite base stations without needing any further physical lines.

The Airnet SDR design performs an analog downconversion to a com-mon intermediate frequency followed by digitization. The design can digi-tize a 5-MHz-wide band of spectrum [17, 18] and can therefore include many GSM carriers. A channelizer made up of DSP circuitry isolates the

GSM carriers (200 kHz) from the wideband digital stream. The reverse process is conducted for the transmit channels using a DSP combiner, which can add several narrowband signals for transmission as a single 5-MHz-wide stream via a single linear power amplifier. The combination of the channelizer and combiner is known as a Carney Engine, and the baseband processing unit in a BTS consists of a Carney Engine and the baseband DSP circuitry. Using terms from this book, the Carney Engine is functionally equivalent to digital frequency up and downconversion.

The Carney Engine can be configured to channelize (isolate and convert to baseband) every second GSM channel within a 5-MHz band. The baseband channels can be routed to virtually any one of the baseband DSPs within a very generic DSP architecture. This approach allows up to 12 GSM channels from the Carney Engine to be baseband processed and produce as many as 96 duplex traffic channels. The architecture is designed so that software can be developed for the BTS to allow processing of different layer one protocols (e.g., GSM, TACS, and NMT900). The design also supports multiple simultaneous air interfaces, and Airnet has demonstrated this for TACS and NMT900.

8.3.3 3G SDR Testbeds

There have been many software defined radio projects over the last several years, and in this section we look at two.

8.3.3.1 SDR Base Station Testbed

The French company Alcatel is known for developing both infrastructure and terminal equipment for the cellular mobile phone industry. Members of the company's research and innovation group have developed a software defined radio base station testbed [19]; a block diagram of the system is presented in Figure 8.34. This system aims to demonstrate the feasibility of a flexible, modular multiband and multimode radio system and provide SDR solutions that can be applied to future base station products.

The testbed's analog front-end (AFE) interfaces receive and transmit antennas at the RF interface and ADCs and DACs at the digitizing interface. The AFE is based on standard dedicated analog front ends, which convert the RF signal to a common digitally sampled intermediate frequency (IF). The AFE, therefore, incorporates all the necessary amplification, frequency conversion, and filtering prior to analog to digital and digital to analog conversion stage. The digital front end (DFE) consists of digital filtering, chan-

Analog front end | A/D-D/A conversion | Digital front end | Baseband processing

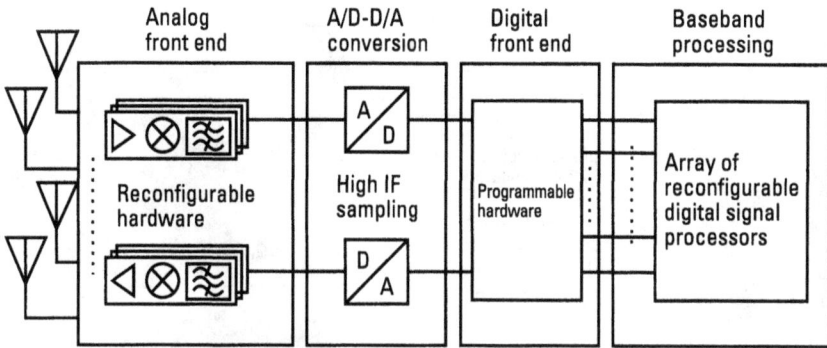

Figure 8.34 SDR base station testbed. (*Source:* Alcatel, 2002. Reprinted with permission.)

nelization, data rate adaption, and digital frequency up and downconversion. Baseband processing (BB) performs functions such as modulation, channel encoding, framing, and the reciprocal receive functions; the BB output is a stream of raw binary data.

The SDR base station testbed has been designed to implement the most critical parts of a multistandard BTS that is able to handle UMTS, GSM, and DCS 1800. The system achieves this goal by combining real-time signal processing and off-line signal evaluation. The testbed can be used to evaluate new algorithms important in the area of SDR, such as power amplifier linearization, multicarrier applications, and digital frequency up and downconversion. The hardware implementation of the radio is depicted in Figure 8.35.

The testbed enclosure includes five major cards: analog transmitter, programmable local oscillator, analog receiver, and baseband processing, with the fifth card hosting the ADC, DAC, and DFE functions. The baseband processing card uses a single Chameleon Reconfigurable Communications Processor RCP (see Section 6.4.1).

8.3.3.2 UMTS TDD Software Radio Testbed

Although this chapter has focused on the FDD variant of UMTS/WCDMA, the time division duplex (TDD) mode is just as applicable for software radio applications. A software radio project being undertaken by Institut Eurecom and Ecole Polytechnique Federale de Lausanne has yielded a UMTS TDD software radio testbed [20]. Compared with FDD, TDD has the advantage that it uses only a single radio channel for both uplink and downlink transmission; therefore, apart from temporal variations, the transmit and receive

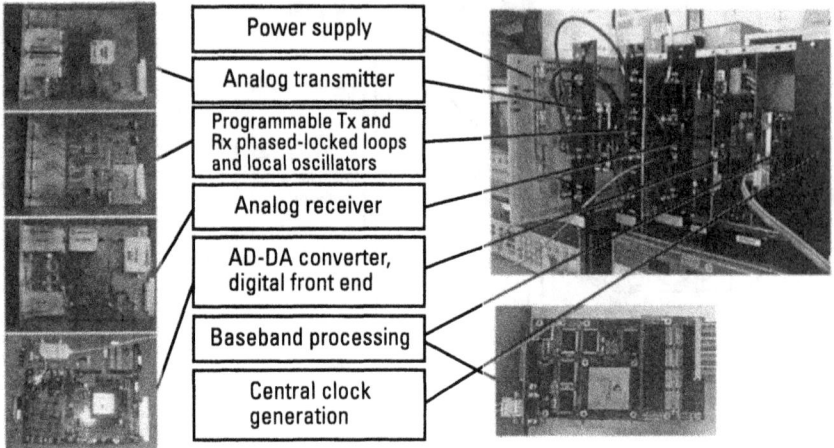

Figure 8.35 SDR testbed hardware implementation. (*Source:* Alcatel, 2002. Reprinted with permission.)

channel response is the same. TDD yields more accuracy when using uplink signal measurements to perform downlink signal processing (e.g., base station transmit beam forming). The TDD testbed is not intended to implement a complete base station; however, Eurecom are using it to validate advanced algorithms in the smart antenna arena as well as for work on multiaccess and power control problems.

Table 8.7
TDD Testbed System Specification

Parameter	Specification
RF Frequency	2.1 GHz
IF Frequency	70 MHz
IF Bandwidth	5 MHz
Chip Rate	3.6864 Mcps
Oversampling Factor (OSF)	4X
IF Sampling Rate	4X 3.6864 MHz = 14.7456 MSPS
Spreading Factor	N = 16
Peak Bit Rate	397.44 Kbps

The testbed hardware architecture uses the cPCI form-factor with commercially available DSP (Spectrum-Barcelona) and radio cards. The remainder of the system consists of a PC and in-house designed ADC, DAC, and PMC cards. The DSP board uses TMS320C6201 200-MHz DSPs. The time duplexing arrangement for TDD divides a 10-ms frame into 15 equal 667-μS slots; these can be used for either uplink or downlink depending on the symmetry required. WCDMA-TDD is a mix of GSM and WCDMA-FDD in that users are both multiplexed in time between slots and multiplexed in the code domain during a time slot. The Eurecom project is configured to use one time slot for uplink and another for downlink with two or three users. (See Table 8.7.)

DSP software was developed using hand-optimized parallel assembly [21]. Reference [21] gives hints on how to optimize code for the C6000 DSP; Chapter 5 deals with optimizing assembly code via linear assembly.

Two DSPs (A and B) are responsible for performing the transmit and receive functions for the two slots and number of users. The utilization of DSP resources for each of these functions is listed in Table 8.8.

Overall, for three users, the implementation consumes 440 μS for the transmit slot and 780 μS for the receive slot.

Table 8.8
DSP Resource Consumption

Function	Clock Cycles	Target Processor
Transmitter		
Spreading and scrambling	7,700 (per slot)	DSPA
Spreading and scrambling	6,300 (per user)	DSPA
Pulse shaping, oversampling, and modulation	61,500 (per slot)	
Receiver		
Primary synchronization code correlation	92,160 (per slot)	DSPB
Joint channel estimation	15,600 (per slot, 3 users)	DSPB
Channel analysis	1,100 (per slot)	DSPB
Matched filter synthesis	5,000 (per user per slot)	DSPB
Passband matched filtering and data detection	44,000 (per user per slot)	DSPB
Total clock cycles	233,360	

8.4 3G Networks

Mobile terminals and base stations are just parts of a complete network that include other major components, such as base station controllers, mobile switching centers, packet gateways, and so on. In this section we cover some of the issues facing mobile network designers and look at an example of a CDMA2000 roll-out.

8.4.1 CDMA2000

The network architecture shown in Figure 8.36 illustrates the major components in a CDMA2000 3G mobile network and their interconnections. Voice and data are communicated between mobiles (phone, PDA, and so on), switched (PSTN, PLMN), and packet-based (Internet) networks. In terms of capacity the architecture suggests that 64 BTSs can be connected to a BSC, and 12 BSCs can be connected to an MSC and GAN; therefore, the core network can support a total of 768 base stations.

The network designer must provide system capacity commensurate with the demand in a given geographic area. The capacity is provided on the basis of predicting a certain quality of service (QoS) (i.e., probability of making a connection, average data rate, or signal to noise ratio) during the connection and the probability of losing an established connection. Mobile TDMA networks are quite unelastic in response to demands for additional

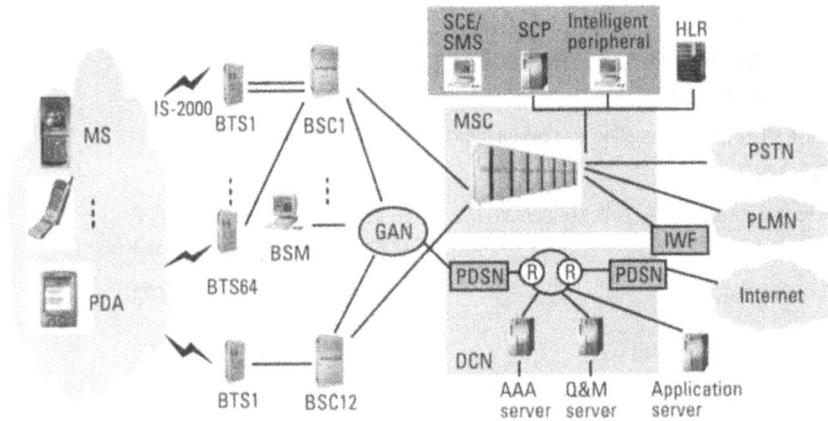

Figure 8.36 CDMA2000 network architecture. (*Source:* Samsung, 2002. Reprinted with permission.)

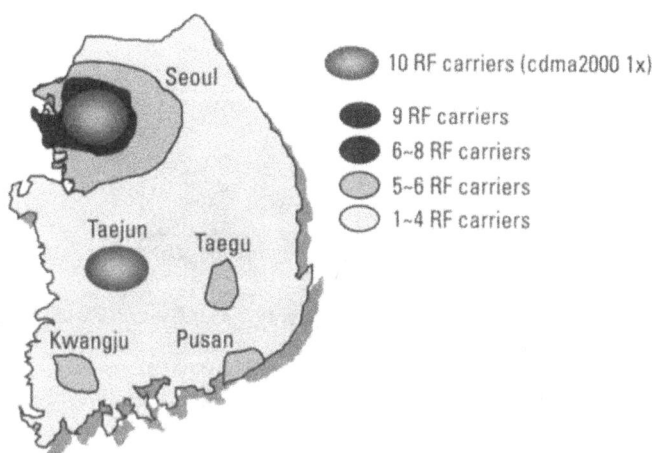

Figure 8.37 Geographic distribution of base station capacity. (*Source:* Samsung, 2002. Reprinted with permission.)

capacity; however, CDMA by its nature is softer in that more users can generally be accommodated at the expense of QoS and cell radius or handover characteristics. Figure 8.37 provides a snapshot of the geographic distribution of 2G and 3G (CDMA2000 1X) capacity for the Korean mobile network operator SK Telecom.

More spectrum and larger-capacity sectored base stations are required for the more densely populated cities, with the outlying and country areas needing only omnidirectional single RF carrier installations. 3G networks will be rolled out gradually, starting in the larger cities, with greater coverage and capacity being added as the demand increases for higher-bandwidth services. Figure 8.38 shows the initial CDMA2000 1X capability for SK Telecom, where two cities are covered by 817 base stations, each with a single RF carrier.

The initial 3G deployment added to the capacity already provided by the 2G IS-95 network, and as of March 2001 the 817 base stations served 45,000 subscribers. Using simple switched circuit theory, if we assume that subscriber/user density is uniform and that during the busiest hour each subscriber generates 0.1 Erlang (requiring service 10% of the hour), each BTS will be servicing $0.1 \times 45,000/817 = 55$ subscribers.

The SK Telecom network was later expanded during four phases to bring the total number of BTSs to 1,737; the voice and data capacity of the network is detailed in Table 8.9.

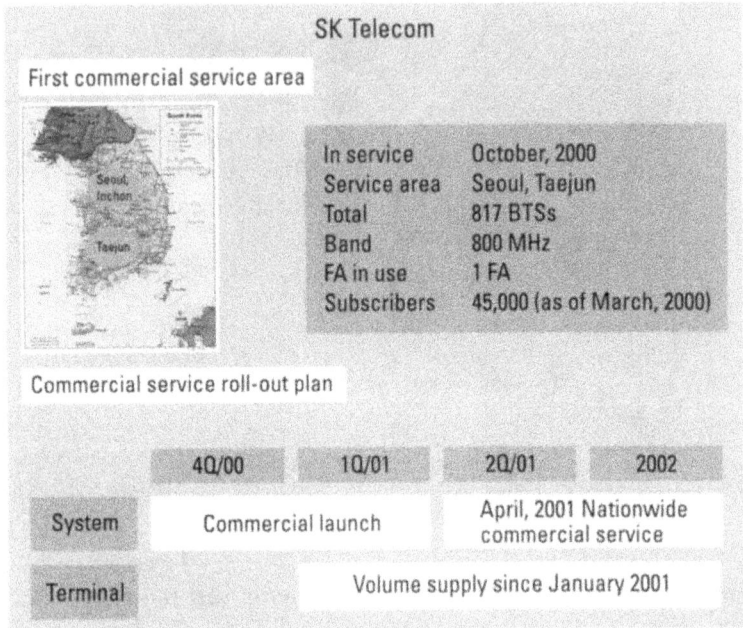

Figure 8.38   An initial CDMA2000 deployment. (*Source:* Samsung, 2002. Reprinted with permission.)

The combined 2G and 3G voice and data capacity of the three-sectored BTS is 624 KB/sec of data plus 111 voice connections; if a sectored BTS is only required to support voice, it has enough capacity for 219 connections. The QoS statistics for the CDMA2000 data component of the SK

Table 8.9
Voice and Data Capacity

IS-95A/B		CDMA2000 1X
Voice Only		
Omni	32 Channels	52 Channels
3 Sector	84 Channels	135 Channels
Voice and Data		
Omni	64K × 1 + 15 Channels	144K × 1 + 43 Channels
3 Sector	64K × 3 + 39 Channels	144K × 3 + 72 Channels

(*Source:* Samsung, 2002. Reprinted with permission.)

Model	SCH-X100
Launch Date	11/15/2000
LCD	7 Line 4Gray LCD
Battery Battery life	520mAh 800mAh Talk: 135 min 190 min Standby: 140 hr 290 hr
Dimension	85 X 45 X 18.8mm
Weight	80g
Key features	Max 144Kbps data service WAP browser Outside LCD Caller ID

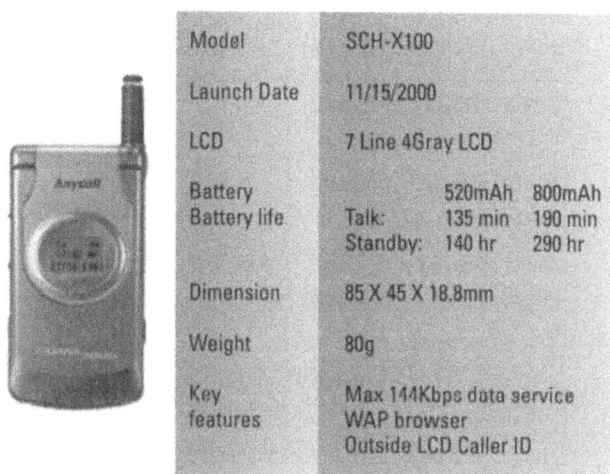

Figure 8.39 CDMA2000 mobile. (*Source:* Samsung, 2002. Reprinted with permission.)

Figure 8.40 CDMA2000 1X indoor BTS. (*Source:* Samsung, 2002. Reprinted with permission.)

Table 8.10
CDMA2000 1X BTS Specification

Supports PCS or cellular bands
Supports 2G
3G capability by S/W upgrade
Supports multisectors: Omni, 2/3 sectors
Carrier capacity per rack: 3 carrier/3 sector, 12 carrier Omni
Channel capacity: 54 Channels/Card
Scalable capacity:
3 carrier/3 sector with 1 rack configuration
6 carrier/3 sector with 2 rack configuration
9 carrier/3 sector with 3 rack configuration
Software-based system expansion:
Channel elements increment by groups of 6
Provides 20W per carrier
Hot-swappable modules
Built-in test unit
Separate battery backup unit
Physical size: 1,800 × 800 × 700 (H × W × D) mm

Source: Samsung, 2002. Reprinted with permission.

Telecom network were: a call establishment success rate of 98%, a call drop rate of 1%, and an average throughput of 105 Kbps.

An example of the CDMA2000 mobiles and base stations used for this Korean network area illustrated in Figures 8.39 and 8.40, respectively, and the BTS specification is listed in Table 8.10.

8.4.2 Other 3G Networks

Other 3G systems are expected to be expanded in similar ways. GSM dominates the European 2G landscape, and it is widely expected that WCDMA services will start appearing as small "islands" of 3G coverage in densely populated areas and then spread out to the suburban regions; this prediction is known as "islands of UMTS in a sea of GSM."

8.5 Conclusion

The initial deployments of 3G have been hardware radio–based and either experimental or done only on a regional basis. The first deployments of UMTS have needed to solve significant interoperability issues and may have required hardware changes to effect a working system. The downturn in the telecommunications sector has placed significant pressure on cost, and this is an area where software radio can provide advantages for system designers by providing a common platform across multiple standards.

References

[1] ITU, "Recommendation ITU-R M.1225 Guidelines for the Evaluation of Radio Transmission Technologies for IMT-2000," 1997.

[2] ETSI, "Submission of Proposed Radio Transmission Technologies," January 29, 1998.

[3] http://www.3gpp.org.

[4] ETSI, "ETSI TS 125.201 Universal Mobile Telecommunications System (UMTS): Physical Layer—General Description," V 3.1.0, June 2000.

[5] ETSI, "ETSI TS 125.213 Universal Mobile Telecommunications System (UMTS): Spreading and Modulation (FDD)," V 4.2.0, December 2001.

[6] ETSI, "ETSI TS 125 211 Universal Mobile Telecommunications System (UMTS): Physical Channels and Mapping of Transport Channels onto Physical Channels (FDD)," V 4.3.0, December 2001.

[7] ETSI, "ETSI TS 145.001 Digital Cellular Telecommunications System (Phase 2+): Physical Layer on the Radio Path (General Description)," V 4.1.0, January 2002.

[8] ETSI, "ETSI TS 125.212 Universal Mobile Telecommunications System (UMTS): Multiplexing and Channel Coding (FDD)," V 4.3.0, December 2001.

[9] TIA, "The CDMA 2000 ITU-RTT Candidate Submission (0.18)," July 27, 1998.

[10] 3GPP2, "C.S0001 Introduction to CDMA 2000 Spread Spectrum Systems," Version 3.0, June 2000. For further information on how to obtain 3GPP2 Technical Specifications and Technical Reports, please visit http://www.3gpp2.org or contact its publications coordinator at 703-907-7088.

[11] 3GPP2, "C.S0002 Physical Layer Standard for CDMA 2000 Spread Spectrum Systems," Version 3.0, June 2001. For further information on how to obtain 3GPP2 Technical Specifications and Technical Reports, please visit http://www.3gpp2.org or contact its publications coordinator at 703-907-7088.

[12] Lee, J. S., and L. E. Miller, *CDMA Systems Engineering Handbook*, Norwood, MA: Artech House, 1998, p. 817.

[13] ITU, "Guidelines for Evaluation of Radio Transmission Technologies for IMT-2000," ITU-R M.1225, 1997.

[14] Software Defined Radio Forum, "Distributed Object Computing Software Radio Architecture," v1.1, July 2, 1999.

[15] Joint Tactical Radio System (JTRS) Joint Program Office, "Software Communications Architecture Specification MSRC-5000SCA," V2.2, November 17, 2001.

[16] Lucent, "Flexent OneBTS NodeB Brochure," FLEXONEBTS, April 2001.

[17] Doner, J. R., "Designing a Broadband Software Radio," *Communications Design Magazine*, http://www.commsdesign.com/main/feat9611.htm, November 1996.

[18] Vu, T., "SDR Success Story in 2.5G and 3G Implementation," http://www.sdrforum.org, June 14, 2001.

[19] Rouffet, D., and W. Konig, "Software Defined Radio," *Alcatel Telecommunications Review*, 2001.

[20] Bonnet, C., et al., "A Software Radio Testbed for UMTS TDD Systems," http://www.eurecom.fr/Sradio/ISTMobSum00.doc, October 2000.

[21] Texas Instruments, *TMSC6000 Programmers Guide*, SPRU198F, February 2001.

9

Smart Antennas Using Software Radio

Smart antenna technology is beginning to make its way into the commercial market, with several companies offering software and hardware for inclusion in mobile cellular base stations. The technology is generating significant interest, because it offers the possibility to reduce the number of base station sites, improve system capacity, and/or reduce spectrum utilization. The renowned Andrew Viterbi [1] is quoted as follows: Spatial processing remains as the most promising, if not the last, frontier in the evolution of multiple access systems.

9.1 Introduction

In this chapter, we introduce the concepts of beam forming and smart antenna technology. We explore several architectures and smart antenna algorithms that can be implemented using software radio. Smart antenna systems don't rely on software radio, but their maximum potential is realized when software control is possible and design is implemented as part of an integrated digital radio architecture.

9.2 What Is a Smart Antenna?

First- and second-generation mobile cellular base stations spread radiation over the entire area of a cell, regardless of the location of the user. Figure 9.1

Figure 9.1   Typical 2G basic antenna system.

depicts a standard GSM base station antenna system for a single directional sector (pointing down a ski run); two panel antennas are used for diversity reception, with the third panel dedicated for transmission.

Smart antenna technology offers the possibility of carefully controlling radiation and reducing exposure. For example, it has been shown [2] that base station power output for a CDMA2000 system can be reduced by approximately 15 dB from 20W using standard antennas to 5.9W when employing a smart antenna system. With increasing scientific and community concern over radiation hazards, smart antenna technology may become a mandatory element in controlling RF power levels in the environment.

Some smart antenna systems are designed for the retrofit market and are not integrated into the radio architecture. When applied to a legacy transceiver system, this solution requires a significant integration effort; this may not be economical if the smart antenna upgrade is required to provide additional system capacity.

The SDR Forum [3] definition for a smart antenna is: A subsystem, which includes the antenna (and possible other classes), that uses the spatial domain in combination with decision-based signal processing to improve link performance and enable other value-added services. It consists of both the software and the hardware objects associated with the additional processing capability.

In a terrestrial digital mobile cellular system important link and system parameters include carrier to interference ratios, cell reuse patterns, number of users per square kilometer, and blocking probability (e.g., see Figure 16 of [4]).

9.3 3G Smart Antenna Requirements

The CDMA2000 system is one 3G standard that has added functionality for smart antennas; the auxiliary pilot channel [5] is described as an unmodulated, direct sequence spread spectrum signal transmitted continuously by a CDMA base station. An auxiliary pilot channel is required for forward link spot beam and antenna beam-forming applications and provides a phase reference for coherent demodulation of those forward link CDMA channels associated with the auxiliary pilot. Similarly, the WCDMA UMTS standard proposes that the secondary common pilot channel (CPICH) and the secondary common control physical channel (CCPCH) forward access channel part can be transmitted in only parts of the cell, therefore requiring downlink beam forming.

9.4 Phased Antenna Array Theory

Before the advent of digital signal processing, phased antenna arrays were designed using two or more antenna elements with a transmission line compensation and combination network implemented in hardware. Figure 9.2 illustrates an n element antenna array, where gain relative to a single element is proportional to the aperture of $(n - 1)d$ m. The array has $n - 1$ degrees of freedom and therefore allows the same number of nulls and peaks to be inserted at arbitrary azimuth angles in the radiation pattern.

Phased array theory is completely reciprocal. If Figure 9.2 is considered to represent a receive system, it shows a planar wave hitting the antenna array at an angle, θ; this will cause a progressive phase shift or time delay in the signals arriving at each antenna element. The transmission line network provides the necessary phase corrections to the signals at each element, so when they are combined a static beam is formed in the spatial domain. This is a narrowband beam former, because the fixed phase corrections imply that the beam shape will change with frequency.

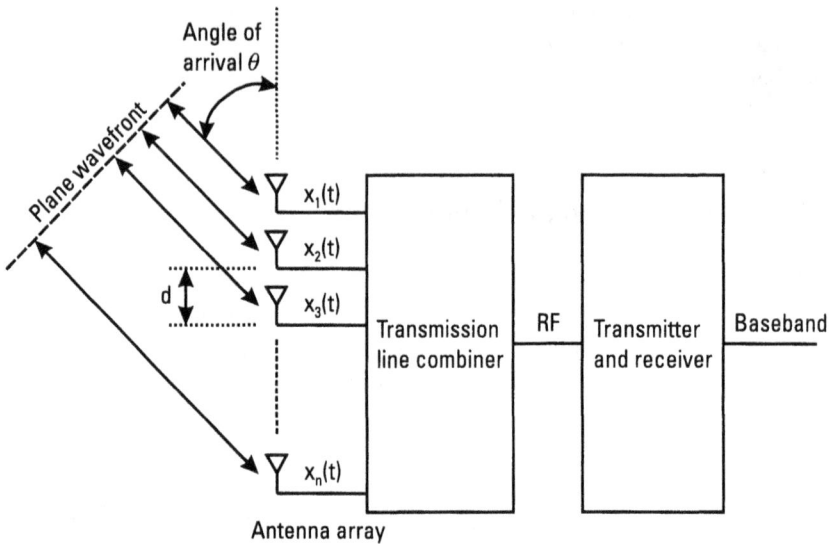

Figure 9.2   Traditional transmission line compensated antenna array.

9.5   Applying Software Radio Principles to Antenna Systems

Even though digital signal processing technology is pushing the software radio closer to the antenna terminals, it has a long way to go before emulating the analog transmission line network in a phased antenna array for frequencies above 100 MHz. Thankfully this isn't necessary, because digital signal processing is able to perform the equivalent radio frequency function at baseband using quadrature processing.

The basic SDR architecture (see Figure 2.3) is expanded in Figure 9.3 for a smart antenna application. The bidirectional data flow arrows indicate the reciprocal nature of the data processing functions (i.e., the system could be a transmitter or receiver).

A wideband analog front end is required for each antenna element; for receive processing it converts from analog RF to a digital IF and vice versa for transmit processing.

Each digital IF channel is fed into a pool of digital frequency conversion and baseband processing resources. The resource pool is responsible for converting the digital IF to quadrature (I and Q) baseband and then performing the smart antenna, modulation/demodulation, and channel coding/decoding functions.

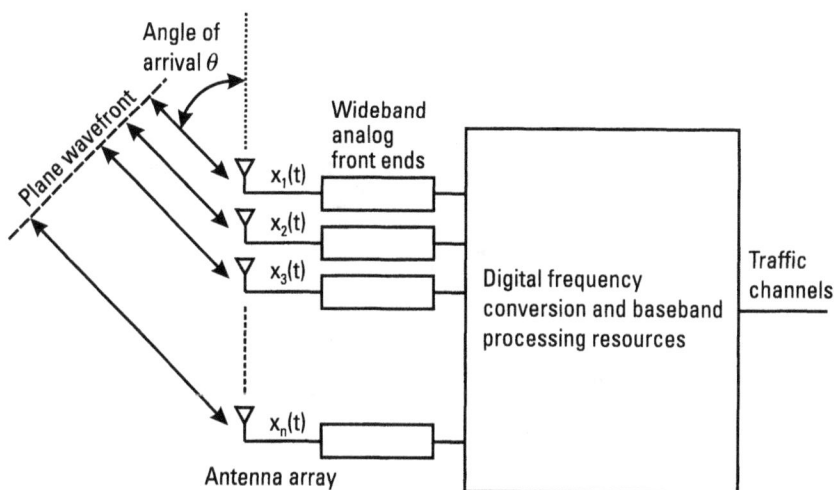

Figure 9.3 Smart antenna system using software radio.

The architecture relies on maintaining phase alignment between each chain from the antenna to the input of the smart antenna function. Because this architecture multiples by n the number of analog front ends and increases the size of the required processing pool, it is mostly suitable for high traffic capacity systems where spectrum is limited. For this case it can become cost effective to expand the capacity of a base station by adding a smart antenna function rather than purchasing more spectrum.

System capacity can also be expanded by cell splitting (i.e., macrocells are divided into many more microcells). This requires more BTS sites, and in situations where it is impossible to operate more sites (e.g., community opposition) the only option may be to expand existing site capacity by the addition of a smart antenna capability.

9.6   Smart Antenna Architectures

There are three major smart antenna architectures: switched beam, optimum combining/adaptive, and direction of arrival (DOA). The major differences relate to the control processes and algorithms applied in the digital signal processing stage. The array of antenna elements can be considered separately, and theoretically each architecture could make use of the same antenna array. The number and physical configuration of antenna elements depend upon the tar-

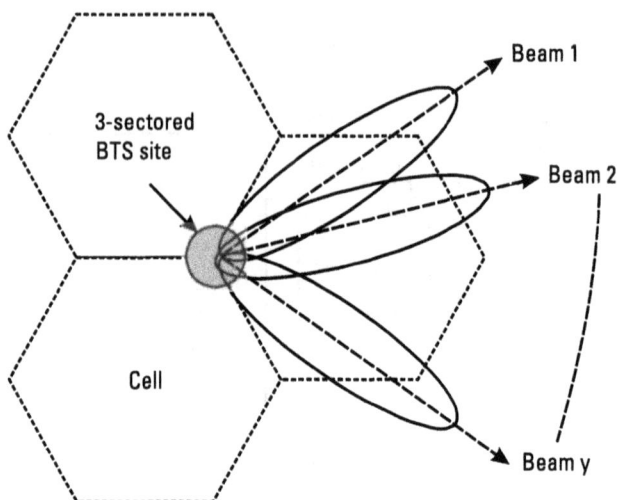

Figure 9.4   Switched beam smart antenna coverage.

get system performance, with improvement generally diminishing as more elements are added (e.g., Figure 13 of [4]). Prototype and commercial systems using between 3 and 12 elements have proven worthwhile and successful.

9.6.1   Switched Beam Arrays

This architecture forms y fixed and overlapping beams to provide a complete azimuth coverage; see Figure 9.4.

Considering the architecture of Figure 9.3 and a receiver system, the processing pool has the capability to process the multiple quadrature baseband signals and simultaneously form multiple wideband beams that each point in a different fixed direction.

The beams can be formed by applying fixed phase corrections to each quadrature channel and summing them prior to the demodulator. The beam-former architecture shown in Figure 9.5 illustrates a mapping of the beam-forming algorithm onto the processing pool. The complex baseband signals, $s_1(t)$, $s_2(t)$, . . . , $s_n(t)$, are multiplied by an array of complex weights, W_1, W_2, . . . , W_n; the results are added to form the quadrature output $v(t)$.

The kth complex weight W_k:

$$W_k = a_k e^{j \sin \theta_k} \tag{9.1}$$

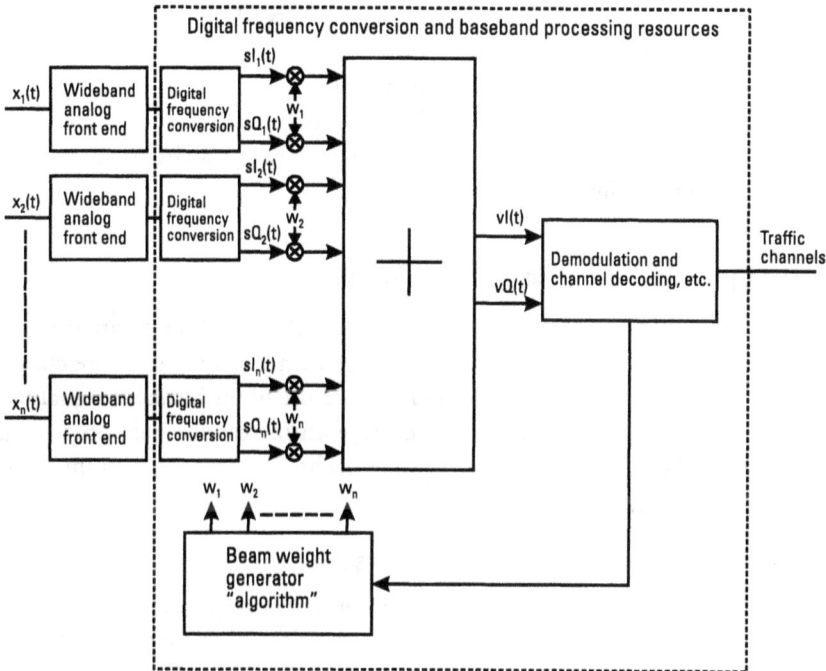

Figure 9.5 Basic beam-former architecture.

where a_k is the amplitude scaling factor for the kth element and θ_k is the phase shift for the kth element.

Expressing W_k in quadrature form:

$$W_k = a_k \cos\vartheta_k + ja_k \sin\vartheta_k \tag{9.2}$$

Expressing the quadrature baseband signal in complex form:

$$s_k(t) = x_k(t) + jy_k(t) \tag{9.3}$$

The input to the complex adder becomes:

$$s_k(t)W_k =$$
$$a_k\left(\left(x_k(t)\cos\theta_k - y_k(t)\sin\theta_k\right) + j\left(x_k(t)\sin\theta_k + y_k(t)\cos\theta_k\right)\right) \tag{9.4}$$

A major advantage with this architecture is its simplicity and relatively lower requirements for signal processing power. Also, for transmission systems where the up- and downlink frequency is separated by a substantial percentage bandwidth, this system offers the ability to closely match the transmit and receive beams. The main limitation with a switched beam system is its ability to limit interference.

9.6.2   Optimum Combining/Adaptive Arrays

The basic hardware architecture of Figure 9.3 applies to all the smart antenna designs in this chapter. This basic architecture with multiple hardware chains increases cost, so it seems logical to make more use of the available signals by matching the bigger hardware architecture with increased signal processing capacity and implementing more complex techniques (e.g., optimum combining and adaptive algorithms).

9.6.2.1   Optimum Combining/Adaptive Algorithms

Algorithms can be used to extract a system output; compare it with a known criterion; and. based on the differences/similarities, calculate a set of beamforming weights. Optimal techniques minimize a cost function that is inversely related to the quality of the chosen system output. They evaluate a set of weights that apply to an ensemble of samples (e.g., MMSE, where the mean square error is minimized for the complete set of samples).

A mean solution applied over a large sample set in a radio environment can suffer from large instantaneous errors, and for these cases it is better to apply an adaptive algorithm that iteratively updates the weights by using previously calculated values as part of the new set.

A disadvantage with MMSE algorithms is that the measured criteria need to be known to both transmitter and receiver (e.g., the GSM training sequence). Errors can occur in the presence of cochannel interference, since the GSM training sequence is not unique for each user. This problem is tackled by using a class of blind adaptive algorithms that does not require a training sequence and operates by restoring some known property of the received signal (e.g., some air interface signals can be correlated with frequency shifted versions of themselves).

Since the CDMA air interface is simultaneously shared by many users, adaptive beam-forming algorithms should be able to separate all users blindly. IS-95 CDMA has a distinct advantage over GSM, because every

Table 9.1
Beam-Forming Algorithms

Algorithm Class	Example Algorithms			
Statistically Optimum Beam forming	Minimum Mean Square Error (MMSE)	Maximum SNR	Linearly Constrained Minimum Variance (LCMV)	
Adaptive	Least Mean Square	Recursive Least Squares	Bussgang	
Blind Adaptive				
Least Squares Constant Modulus				
CDMA Adaptive	Multitarget Least Squares Constant Modulus	Gram-Schmidt Orthogonalization	Phase Ambiguity	Sorting Procedure
	Multitarget Decision Directed	Least Squares Despread Respread Multitarget Array (LSDRMTA)	LSDRMTA Constant Modulus	

mobile in the world is assigned a unique spreading code. This property is used in the LSDRMTA algorithm, where the beam-former output is despread by the user's code and then respread as an input to the weight calculation process.

Many beam-forming algorithms have been developed to suit various air interfaces, and Table 9.1 summarizes those presented in [6].

9.6.3 DOA Arrays

The beam-forming approaches presented so far do not necessarily produce a calculation of the DOA of the user's signal. Also, the resultant vector of weights may not produce a beam lobe in the direction of the user (e.g., it may be more beneficial to null an interferer [or noise] than reinforce a user's signal). DOA algorithms are important, since they can be used to support user location services (e.g., by triangulation by two or more base stations) and for estimation in downlink beam-forming systems.

Table 9.2
DOA Beam-forming Algorithms

Algorithm Class	Example Algorithms			
Conventional DOA	Delay and Sum Method	Capon's Minimum Variance Method		
Subspace DOA	Multiple Signal Classification (MUSIC)	Root MUSIC	Cyclic MUSIC	Estimation of Signal Parameters via Rotational Invariance Techniques (ESPRIT)
Maximum Likelihood DOA				
Coherent Signal DOA	Spatial Smoothing	Multidimensional MUSIC		
Iterative Least Squares Projection-Based CMA DOA				
Integrated Approach DOA				

9.6.3.1 DOA Algorithms

Table 9.2 summarizes [6] several DOA algorithms; the integrated approach is particularly useful for CDMA systems.

9.6.4 Beam Forming for CDMA

The Rake receiver (see Figure 8.3) is traditionally only connected to a single antenna element at any given time and is, therefore, unaware of the direction of arrival of each ray. Such a receiver cannot resolve between two or more rays arriving at the same time delay but from different azimuth or elevation angles. Incorporating beam forming into a Rake receiver is of most interest to 3G designers because the UMTS and CDMA2000 air interfaces use CDMA.

A useful smart antenna system for a CDMA receiver combines the Rake principle described in Chapter 8 with a multielement antenna array. This is known as space-time (or two-dimensional) processing and is recommended [7] as suitable for micro- and macrocell applications.

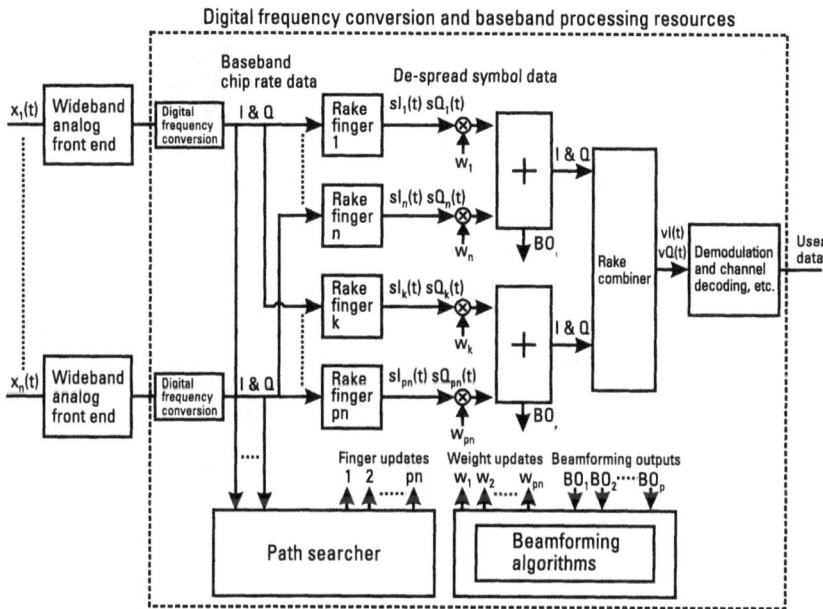

Figure 9.6 Option 1 CDMA beam-forming Rake.

Figure 9.6 illustrates one option, where the functional block diagram of Figure 9.5 is extended to illustrate the concept for a single user. For this first option, Rake fingers are inserted in-between the digital frequency down-conversion stage and the weight multiplication stage. In this way individual beams can be formed for each resolvable multipath prior to the Rake combiner. Spatial processing is not disturbed by the despreading process, since "signals from all elements of the array can be multiplied by a scalar value without distorting the spatial characteristics of the received vector" [8].

There are n antenna chains (analog front end plus frequency conversion) and p beam formers, one for each multipath component. Beams are formed in the direction of the multipath component; this enhances the needed user energy and reduces interference generated by other users (at directions away from the user). The outputs of the p beam formers are fed into a normal Rake combiner [e.g., maximal ratio combiner (MRC)].

The path searcher periodically sends timing updates to each of the Rake fingers, as would normally be the case. New weights are periodically calculated for the p beam formers by an adaptive algorithm. This algorithm can use a system output (e.g., uplink pilot channel data) as its input to measure the quality of the decision or use a blind technique. The period between weight updates will be a function of angular velocity and will, therefore, be impacted by the user's speed and range from the base station.

This option has the advantage of performing space processing (beam forming) at symbol rate but needs $n \times p$ fingers and $n \times p$ path searchers.

The second option, as illustrated in Figure 9.7, is to beam form at chip rate and then despread prior to Rake combining. Table 9.3 summarizes the functional requirements for the two options, where the multipliers are those used in beam forming.

One can estimate the algorithmic load added by these two smart antenna options by considering the UMTS case, where $n = 4$, $p = 4$, spreading factor SF = 8 (user data rate approximately 384 Kbps), OSF = 4 and chip rate = 3.84 Mcps; this analysis neglects the processing load of the weight update algorithms and assumes an addition or multiplication can be completed in one processor cycle. (See Table 9.4.)

Table 9.5 uses data from Table 6.1.

Using the data from Table 6.1 and comparing the signal processing load for the chip-rate receive functions of path searching and Rake (1,500 + 650 = 2,150 MIPS), the option 1 smart antenna solution increases the processing load by 12 times and option 2 increases the load by only 3 times. Note that inclusion of the LS-DRMTA algorithm in the weight update function will significantly expand these figures, since it requires both despreading and respreading of the user's signal.

9.6.5 Downlink Beam Forming

Until now we have assumed reciprocal receive and transmit processing functions; this is a major simplification for the majority of mobile systems. A common transmitter technique uses the receiver's DOA estimate to form the transmitter beams; this can be successful for simple systems such as UMTS-TDD but will lead to inaccuracies in duplex split-band systems such as GSM and CDMA2000.

The downlink contains a number of control channels (e.g., pilot, broadcast, paging, sync) that are accessed by users anywhere in the cell without any communication with the base station. Therefore, any downlink beam-forming scheme must deal with uniform transmission of control channels while simultaneously steering traffic channels to discrete users. Problems can occur for IS-95 when the mobile uses a different path for a control and traffic channel, since the phase, timing, and amplitude can be

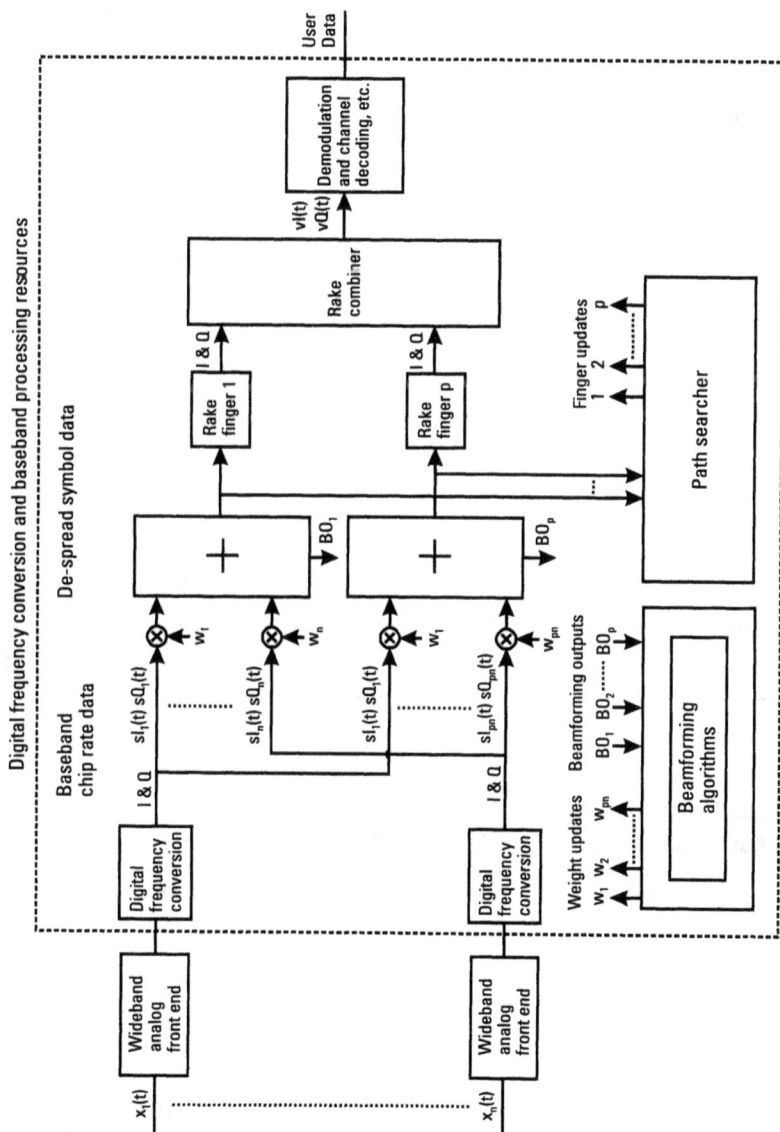

Figure 9.7   Option 2 CDMA beam-forming Rake.

Table 9.3
CDMA Beam-Forming Rake Options Summary

Number of:	Option 1	Option 2
Fingers	$n \times p$	p
Chip-Rate Multipliers	—	$n \times p$
Symbol-Rate Multipliers	$n \times p$	—
Chip-Rate Adders	—	p
Symbol-Rate Adders	p	—
Path Searchers	$n \times p$	p

Table 9.4
CDMA Beam-Forming Rake Intermediate Calculations

Number of:	Option 1	Option 2
Fingers	16	4
Chip-Rate Multipliers	—	$16 \times OSF \times 3.84$ Mcps
Symbol-Rate Multipliers	16×3.84 Mcps / SF	—
Chip-Rate Additions	—	$4 \times OSF \times 3.84$ Mcps
Symbol-Rate Additions	4×3.84 Mcps / SF	—
Path Searchers	16	4

Table 9.5
CDMA Beam-Forming Rake Final MIPS

Number of:	Option 1	Option 2
Finger Processing MIPS	$16 \times 650/4 = 2,600$	$4 \times 650/4 = 650$
Chip-Rate Multiply MIPS	—	245
Symbol-Rate Multiply MIPS	7.7	—
Chip-Rate Add MIPS	—	61.4
Symbol-Rate Add MIPS	1.9	—
Path Searching MIPS	$16 \times 1,500 = 24,000$	$4 \times 1,500 = 6,000$
Total MIPS	26,610	6,956

different. This issue is being addressed by CDMA2000 with auxiliary pilots.

9.6.6 A Software Radio Smart Antenna Architecture

The target physical architecture should include sufficient flexibility so that it can be configured between a system processing more users without a smart antenna capability, through to one with less users and a smart antenna capability. Configuration would be via software, and the architecture should also be easily expandable to increase user capacity with the smart antenna capability. Figure 9.8 suggests a flexible architecture suitable for the purpose.

In general the architecture should allow any wideband analog channel to connect to multiple digital frequency conversion stages; this is because a wideband chain carries more than one RF carrier, which, in turn, contains many users. Similarly, each digital frequency conversion output should be capable of connecting to one or more Rake fingers, since each chip-rate baseband signal carries more than one user. The first switch in the system allows data from each wideband front end to connect to one or many digital frequency converters. The second switch supports a many-to-many data transfer mechanism between digital conversion and the signal processing resources. By implementing the switches in FPGA the hardware becomes software programmable and flexible.

The hardware architecture will need to provide adequate interprocessor, intercard, and interchassis communications to allow the dynamic routing of data so that various software functions can be partitioned across a number of signal processing resources. As an example, the software for the beam weights generator is likely to be partitioned to a single signal processing resource (e.g., DSP); the weight coefficients can be shared with other functions via the central memory pool or with adjacent processors via the interprocessor communications ability.

9.7 Smart Antenna Performance

Using a simulated network [5] and a four-element antenna array, it has been shown that the capacity of a CDMA network with power control can be dramatically increased. For this example the network capacity increased by between four and five times when compared with the same system without a smart antenna capability.

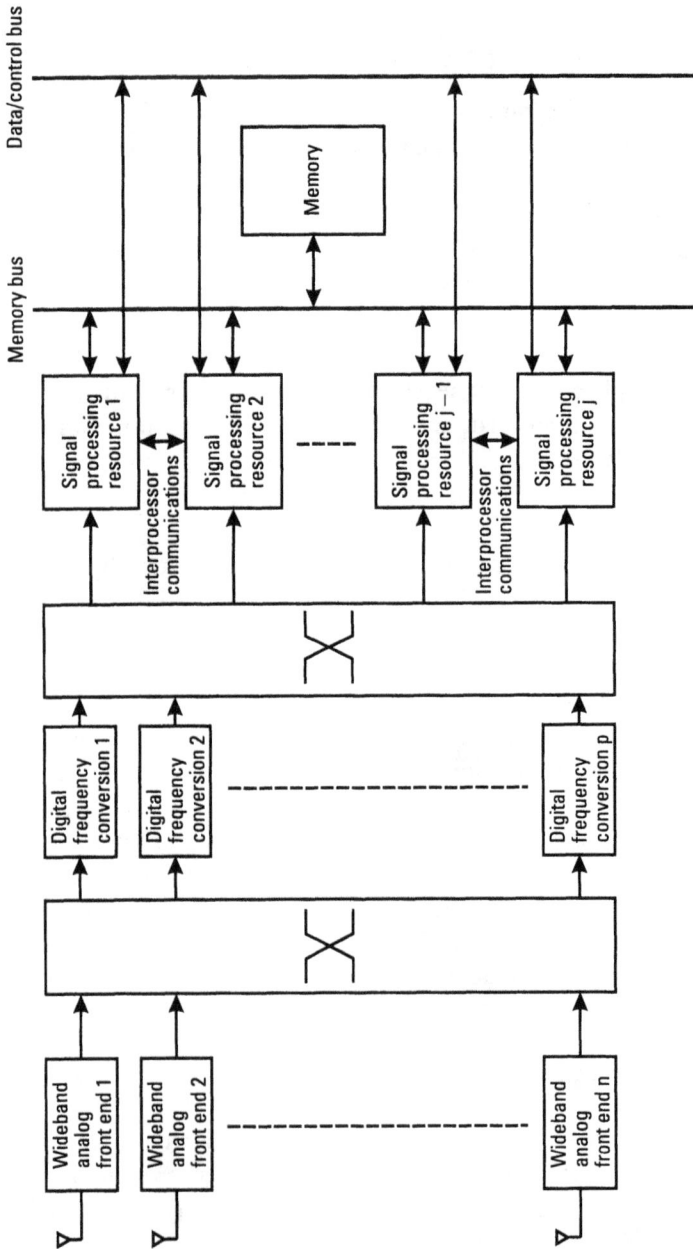

Figure 9.8   A flexible physical architecture.

Another example [9] for a direct sequence CDMA receiver measured the bit error rate (BER) for 16 active users under conditions of SNR = 10 dB and a three-path Raleigh fading channel. The BER reduced by almost two orders of magnitude from 10^{-1} without beam forming to 10^{-3} with a seven-weight beam former.

At the network level the benefits of a smart antenna system for a 1.9-GHz CDMA2000 deployment have been simulated [2]. The work used the San Francisco area with 5.6 million people and covering 5,200 km^2. In conclusion it indicated that the number of base station sites could be reduced by 70% with a commensurate 30% reduction in capital expenditure.

9.8   Conclusion

Even though smart antenna functionality offers significant advantages, it is not a core requirement of the 3G cellular standards. We have shown that smart antenna functionality can be computationally very expensive and, therefore, the decision to use such a capability will be part of the implementation and system design process. Software radio provides the perfect avenue to accommodate differing requirements and allows network designers to defer implementation in the first instance and then upgrade to a smart antenna solution at a later date.

References

[1]   Boukalov, A., "Standardization and System Integration of Smart Antennas into Wireless Networks," *Helsinki University of Technology Communications Lab ETSI/MESA Meeting*, September 18, 2001.

[2]   ArrayComm, "Smart Antenna Technology for CDMA 2000 Network," July 2001.

[3]   The Software Defined Radio Forum, http://www.sdrforum.org/MTGS/mtg_14_jun99/sadef.doc, 1999.

[4]   Razavilar, J., F. Rashid-Farrokhi, and K. J. Liu, "Software Radio Architecture with Smart Antennas: A Tutorial on Algorithms and Complexity," *IEEE Journal on Selected Areas in Communications*, Vol. 17, April 1999.

[5]   3GPP2, "C.S0002 Physical Layer Standard for CDMA 2000 Spread Spectrum Systems," Version 3.0, June 2001, p. 674.

[6]   Liberti, J., and T. Rappaport, *Smart Antennas for Wireless Communications*, Englewood Cliffs, NJ: Prentice Hall, 1999.

[7] Boukalov, A., and S. Haggman, "System Aspects of Smart-Antenna Technology in Cellular Wireless Communications—An Overview," *IEEE Transactions on Microwave Theory and Techniques*, Vol. 48, June 2000.

[8] Liberti, J., and T. Rappaport, *Smart Antennas for Wireless Communications*, Englewood Cliffs, NJ: Prentice Hall, 1999, p. 120.

[9] Mohamed, N., and J. Dunham, "A Low-Complexity Combined Antenna Array and Interference Cancellation DS-CDMA Receiver in Multipath Fading Channels," *IEEE Journal on Selected Areas in Communications*, Vol. 20, February 2002.

10

Low-Cost Experimental Software Radio Platform

10.1 Introduction

Previous chapters have covered the design issues associated with the development of a software radio, with particular emphasis on the 3G cellular mobile standards. We now provide implementation-level detail for an experimental software radio platform.

10.2 Platform Requirements

The system requirements were to design and implement software radio functionality in a high-level software language by using a low-cost platform. The hardware was required to receive radio frequency signals in the several MHz range, demodulate, and audio amplify them.

10.3 System Architecture

The Texas Instruments' C6701 evaluation module (EVM), THS12082 ADC EVM, and Code Composer source code development tools were chosen as the compliant solution. The C6701 EVM hosts a single TMS320C6701 floating-point DSP, and the card uses a full-length PCI form-factor for installation in a PC. The C6701 is provided with an evalua-

Figure 10.1 Texas Instruments' C6701 DSP EVM.

tion copy of the Code Composer software development tool for the Windows operating system. (See Figure 10.1.)

The THS12082 EVM (see Figure 10.2) is designed either to be used standalone or connected to the C6701 EVM via the TMS320C62x EVM daughterboard interface. It contains a 12-bit ADC, which can be clocked at up to 8 MHz, and, importantly, it also features an on-chip 16-word FIFO. Without a FIFO (e.g., THS1209 ADC) the DSP must read every ADC sample before the next one is written; this usually occurs during an interrupt service routine (ISR) after the ADC has interrupted the DSP indicating that new data is ready. In this situation the DSP needs a fixed amount of time (latency) to halt the processor and enter the ISR to perform the read. This

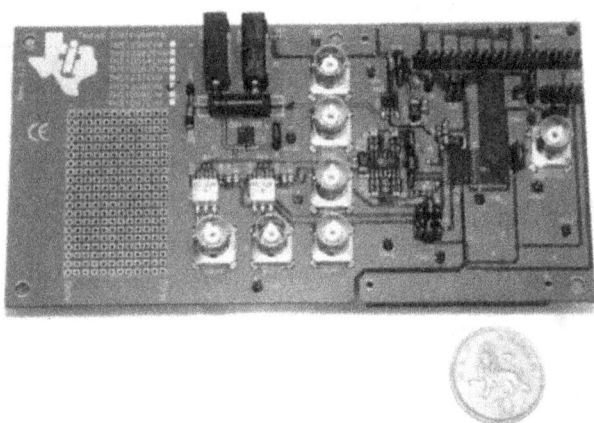

Figure 10.2 Texas Instruments' THS12082 ADC EVM.

latency limits the frequency of DSP reads and reduces the number of cycles available for signal processing. For example, a TMS320C542 clocked at 40 MHz experiences a 400-nanosecond latency [1] and will not be able to keep pace reading samples from a THS12082/THS1206 clocked at 2 MHz. The same processor can reach a 6-MHz transfer rate when the FIFO is used and the trigger level (FIFO depth) is set to eight words. Even though the trigger level can be increased, it has been found [2] that eight is the optimum figure to ensure that old data is not overwritten.

The system hardware architecture is illustrated in Figure 10.3 and shows the major system interfaces and signals.

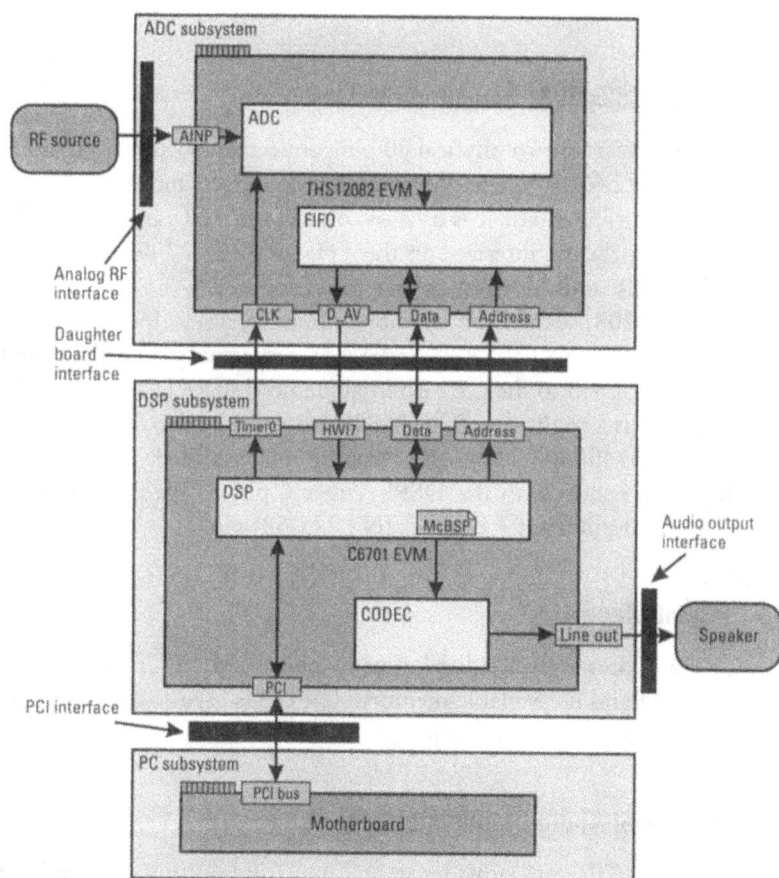

Figure 10.3 System hardware architecture.

10.4   System Interfaces

There are four major system interfaces, including two analog interfaces, a TMS320C62x EVM daughterboard interface, and a standard PCI interface.

10.4.1   Analog RF Interface

As shown in Figure 10.2, the THS12082 EVM contains seven on-board BNC sockets (AINP, ADIFF, AINM, BINP, BDIFF, and BINM [3]) to connect an RF source via coaxial cable. The single ended AINP interface is a suitable choice for connection of an RF test source and for use in the final system.

10.4.2   TMS320C62x EVM Daughterboard Interface

This interface consists of two physical 80 pin connectors (J6 and J7) that provide access to the C6701's expansion peripheral interface and the expansion memory interface. The interface is used to connect the ADC and DSP EVMs together. Importantly, the interface on the THS12082 EVM does not use all of the possible pins, and this must be taken into account in the software.

The THS12082 expansion memory interface, J6-1, J6-4, and J6-3, uses DSP address pins XA17, XA20, and XA21, respectively. In MAP1 mode this memory is mapped to the CE1 external memory space [4], 0x01400000 to 0x016FFFFF; as a result the THS12082 is accessed by the DSP via memory address 0x01400000. For the expansion peripheral interface the THS12082 only connects to the DSP's Timer 0 pin (TOUT0 J7-45) and the DSP's external interrupt 7 (XEXT_INT7 J7-53) pin.

10.4.3   PCI Interface

The Code Composer software development tool uses the PCI bus to access the C6701 EVM and its available memory. The bus is used to debug the system, load code, and run the software.

10.4.4   Line-Level Audio Output Interface

The rear of the C6701 card provides several input and output physical interfaces, including a 3.5-mm stereo audio output interface. Multichannel buffered serial port 0 (McBSP) on the DSP is connected to a CS4231A [5] audio codec on the EVM, which, in turn, is connected to the audio output inter-

Table 10.1
Codec Sampling Frequency

I8 Value	Codec Sampling Frequency f_c
0x40	8.0 kHz
0x42	16.0 kHz
0x44	27.42 kHz
0x46	32.0 kHz
0x48	NA
0x4A	NA
0x4C	48.0 kHz
0x4E	9.6 kHz

face. The codec is capable of producing both mono and stereo line-level audio using a range of sampling rates. Register I8 of the codec can be programmed via the DSP for the sampling frequencies f_c listed in Table 10.1, assuming that the 24.576-MHz on-board crystal is selected, mono is chosen, and linear 16-bit two's complement little endian data format is used.

10.5 System Design

This section covers clock generation and sample rate matching.

10.5.1 DSP Clock Frequency

An on-board oscillator is used to clock the DSP, and the frequency is altered by DIP switches SW2-6 and SW2-7 [6]. Clock frequencies f_d of 25, 33.25, 100, and 133 MHz are possible, and the choice will obviously affect the signal processing capacity of the card (MIPS) and determine the need for sample rate conversion.

10.5.2 ADC Clock Source

The on-chip DSP timers can be programmed in C directly or by GUI access in Code Composer's chip support library (i.e., the DSP/BIOS configuration .cdb file). To create a clock source suitable for clocking the ADC use the settings shown in Table 10.2.

Table 10.2
DSP Timer Settings

Parameter	Setting
Input Clock Source (CLKSRC)	(CPU clock/4)
Clock/Pulse Mode (CP)	Clock mode
Pulse Width (PWID)	One clock cycle
Function of TOUT (FUNC)	Timer output
TOUT Inverter Control (INVOUT)	Uninverted
Timer Operation	Start with reset
Period Value	See (10.1)

Once initialized, the timer is free running and does not require any intervention from the DSP or use any DSP cycles.

The timer frequency f_{adc} is determined by a combination of *Period-Value*, and the DSP clock frequency f_d and can be calculated as follows:

$$f_{adc} = \frac{f_d}{4 \times PeriodValue} \tag{10.1}$$

10.5.3 Matching Sampling Rates

Since the codec sampling frequency, f_c, and the DSP clock frequency, f_d, can only be set to a limited range of discrete values, it is desirable to choose the frequencies where they are related by an integral multiple of D to minimize the signal processing load, as follows:

$$D \times f_c = \frac{f_d}{4 \times PeriodValue} \tag{10.2}$$

Signal processing load can be minimized by avoiding the use of an interpolation function (e.g., 5.3.5 CIC filters) to match the input (ADC) and output (codec) sampling rates. For this case the SDR processing function performs a simple decimate by D to match the data rates; some examples of potential combinations of parameters are provided in Table 10.3.

With the fastest DSP clock frequency of 133 MHz, it is not possible to match the data rates exactly for ADC sampling frequencies in the several

Table 10.3
Data Rate Matching Parameters

f_d	Period Value	f_{adc}	f_c	D	Frequency Error
100 MHz	5	5 MHz	8,000 Hz	625	0 Hz
100 MHz	25	1 MHz	8,000 Hz	125	0 Hz
100 MHz	125	200 kHz	8,000 Hz	25	0 Hz
133 MHz	7	4.75 MHz	7,996.6 Hz	594	0.4 Hz
133 MHz	14	2.375 MHz	7,996.6 Hz	297	0.4 Hz
133 MHz	27	1.23 MHz	7,996.6 Hz	154	0.4 Hz

MHz region. For applications that can tolerate a small frequency error (e.g., voice) and the proportional loss of data, this may be an adequate system solution. However, limiting the choice of sampling frequency and the potential for a relatively large frequency error is unlikely to be acceptable for many systems and will complicate other areas of design. Therefore, the preferred solution is to include an interpolation function; this may even be necessary for designs where the expected frequency error is zero, because the DSP and codec clocks are provided by independent sources and will, therefore, have a frequency error and dissimilar drift characteristic.

10.6 Functional Design

To ensure sufficient data throughput and maximum signal processing capacity, the design uses the ADC's FIFO and DSP's on-chip DMA engines. DMA hardware is a very useful resource, since it can be used to transfer data between local memory and external memory/devices in the background, leaving the DSP free to perform signal processing.

The functional design uses Dma0 to transfer frames (bursts) of data from the ADC FIFO to a double buffer (ADC Buffer A and ADC Buffer B) in DSP RAM. The SDR processing function processes (decimation, filtering, demodulation, interpolation, amplification, gain control, and so on) data from this buffer and places the data into a second double buffer (McBSP Buffer A and McBSP Buffer B). Dma1 is used to transfer frames of data from the second buffer to the McBSP0 serial port connected to the codec. A fixed system processing schedule design is depicted in Figure 10.4.

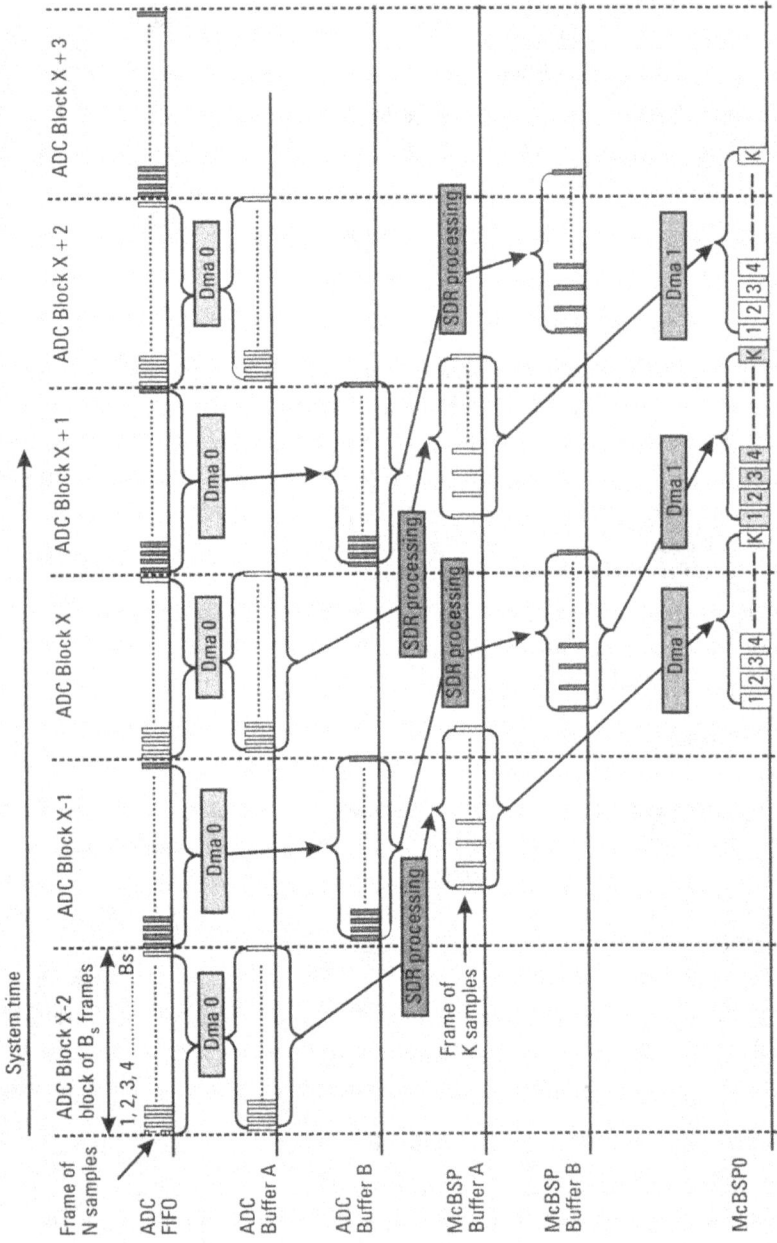

Figure 10.4 System processing schedule.

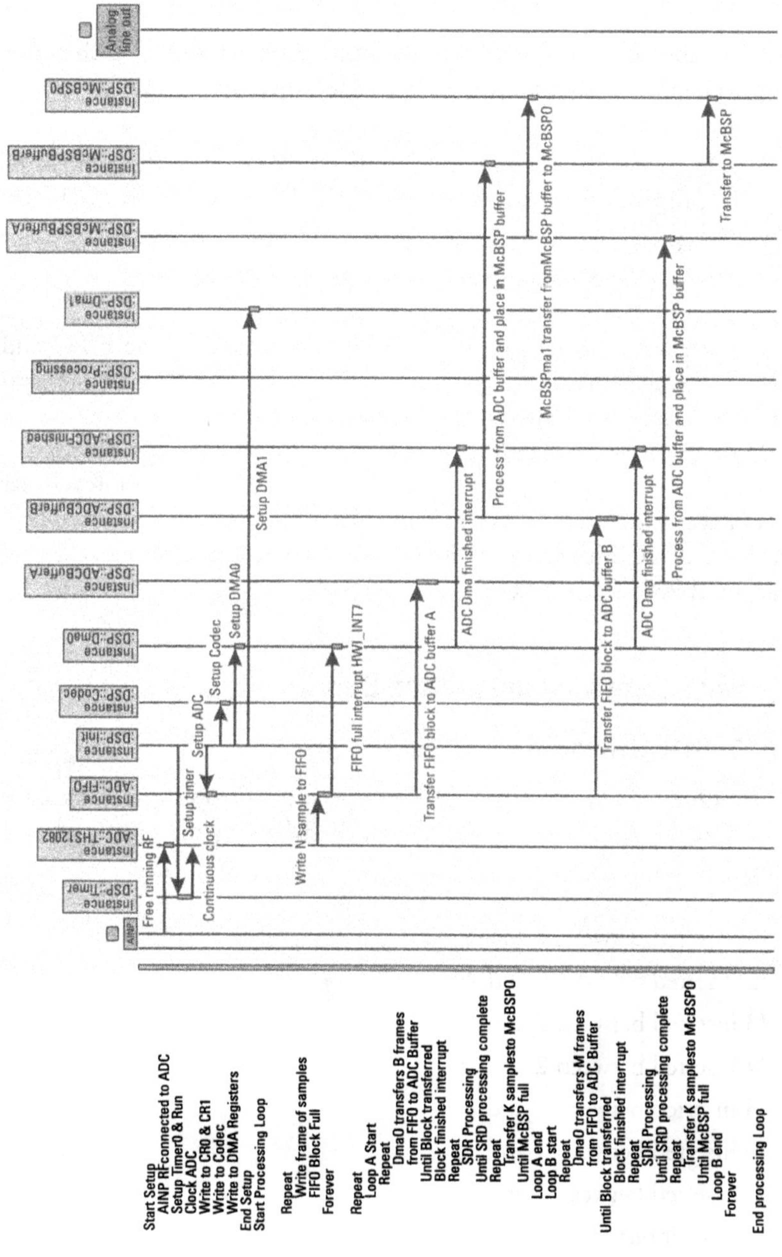
AINP
Instance :DSP::Timer
Instance :ADC::THS12082
Instance :ADC::FIFO
Instance :DSP::Init
Instance :DSP::Codec
Instance :DSP::Dma0
Instance :DSP::ADCBufferA
Instance :DSP::ADCBufferB
Instance :DSP::ADCFinished
Instance :DSP::Processing
Instance :DSP::Dma1
Instance :DSP::McBSPBufferA
Instance :DSP::McBSPBufferB
Instance :DSP::McBSP0
Analog line out
Free running RF
Setup timer
Continuous clock
Setup ADC
Setup Codec
Setup DMA0
Setup DMA1
Write N sample to FIFO
FIFO full interrupt HWI_INT7
Transfer FIFO block to ADC buffer A
ADC Dma finished interrupt
Process from ADC buffer and place in McBSP buffer
McBSPma1 transfer fromMcBSP buffer to McBSPO
Transfer FIFO block to ADC buffer B
ADC Dma finished interrupt
Process from ADC buffer and place in McBSP buffer
Transfer to McBSP
Start Setup
AINP RF-connected to ADC
Setup Timer0 & Run
Clock ADC
Write to CR0 & CR1
Write to Codec
Write to DMA Registers
End Setup
Start Processing Loop
Repeat
Write frame of samples
FIFO Block Full
Forever
Repeat
Loop A Start
Repeat
Dma0 transfers B frames
from FIFO to ADC Buffer
Until Block transferred
Block finished interrupt
Repeat
SDR Processing
Until SRD processing complete
Repeat
Transfer K samplesto McBSPO
Until McBSP full
Loop A end
Loop B start
Repeat
Dma0 transfers M frames
from FIFO to ADC Buffer
Until Block transferred
Block finished interrupt
Repeat
SDR Processing
Until SRD processing complete
Repeat
Transfer K samplesto McBSPO
Until McBSP full
Loop B end
Forever
End processing Loop

Figure 10.5 System sequence diagram.

Considering that the current block of data is ADC Block X, the system processing schedule illustrates that there are three parallel activities, as follows:

1. Dma0 transfers B_s frames of data into ADC Buffer A; each frame consists of N samples, where N is the programmed FIFO depth or trigger level.
2. SDR processing accesses ADC Buffer B (ADC data from Block X-1), processes the data, and places K samples in McBSP Buffer B.
3. Dma1 accesses McBSP Buffer A (ADC data from Block X-2) and transfers the K samples out to the serial port McBSP0.

The schedule is driven by the interrupts generated by the FIFO and mapped to the DSP hardware interrupt HWI_7 and Dma0; therefore, system time is derived from the signal generated by Timer0.

There are software design tools that attempt to capture the same information with a sequence diagram (e.g., see Figure 10.5). This approach has the advantage of including all the system objects but does less well when capturing parallel activities and connecting these to the age of data being processed (e.g., Block X-2, Block X-1, and so on).

10.7   Low-Level Implementation Details

10.7.1   THS12082 Hardware

The THS12082 EVM contains several jumpers; the following settings allow the board to be connected to the C6701 EVM using the daughterboard interface [6]:

J1 inserted between 1 and 2;

J2 inserted between 1 and 2;

J3 inserted between 2 and 5;

J4 inserted between 2 and 5;

J5 inserted between 1 and 2;

J6 inserted between 1 and 2;

J7 inserted between 1 and 2;

J10 not inserted;

J11 is inserted;
J12 inserted between 2 and 3;
J13 inserted between 1 and 2.

10.7.2   THS12082 Software

A data converter support plug-in [7] for Code Composer is available for download [8] from the Texas Instruments' Web site. This plug-in provides an API-level programming interface that generates C code with data structures and configuration parameters based on the state of the values in the GUI. The following code is autogenerated for the THS12082 by the plug-in and provides a structure for access to the ADC via the DSP.

```
TTHS12082 Ths12082_1 = {
    &ths1206_configure,
    &ths1206_power,
    &ths1206_read,
    &ths1206_write,
    &ths1206_rblock,
    &ths1206_wblock,
    0, 0, 0, 0,
    (volatile int *) 0x01400000,
    ADC1_CR0_VALUE,
    ADC1_CR1_VALUE,
    0,
    0,
    0,
    ADC1_TRIGGER_LEVEL,
    ADC1_INTNUM,
    ADC1_SHIFT
};
```

Prior to accessing the ADC, the main program only needs to run the dc_configure function, which sets up the ADC control registers. The dc_read function can be used to read a single sample; however, as discussed, the dc_rblock function is more useful, since it uses DMA to shift a block of data from the ADC into DSP memory.

10.7.3   DSP BIOS Configuration

Code Composer integrates its C development environment with a number of features designed to ease the programming of hardware and abstract low-level details away from the software developer. The chip support library (CSL) and scheduling module are two important parts of DSP BIOS.

The CSL provides high-level programming access to direct memory access (EDMA and DMA), EMIF, McBSP, timer, and expansion bus (XBUS) hardware. Using this feature reduces the amount of C code needed by hiding the data structures that store the various parameters; this approach also reduces the chance of making errors.

The scheduling module provides access to the DSP's 16 hardware interrupts and allows the software developer to create many software interrupts. Overall, the hardware interrupt set has priority over the software interrupt set, and the user can also assign priority levels within each of the sets. A C function can be assigned to each hardware and software interrupt, and in this way the designer is provided with a kernel-level operating system. The software radio can implement a fixed task schedule or make use of DSP/BIOS to schedule and run the tasks.

10.8　Potential Applications and Conclusion

With the addition of a high-gain, low-noise RF receive amplifier prior to the ADC, software could be written for a simple AM or SSB receiver in the MF and low HF broadcasting bands. More demanding applications requiring digital demodulation in the mobile cellular bands could also be implemented with the addition of an analog frequency downconverter prior to the ADC. This would probably require a three-stage downconversion to move from 900 MHz to the 1–6 MHz range. Alternatively, the ADC daughtercard could be replaced with a commercial digital downconverter to reduce the requirement on the number of analog frequency downconversion stages.

References

[1]　Texas Instruments, "Designing with the THS1206 High-Speed Data Converter," SLAA094, April 2000.

[2]　Texas Instruments, "Higher Data Throughput for DSP Analog-to-Digital Converters," August 2000.

[3]　Texas Instruments, "THS1206, THS12082, THS10064, THS10082 Evaluation Module User Guide," SLAU042B, January 2001, p. 4-2.

[4]　Texas Instruments, "TMS320C6201/C6701 Evaluation Module Technical Reference," SPRU305, December1998, pp. 1–12.

[5] Crystal, "CS4231A Parallel Interface, Multimedia Audio Codec," DS139PP2, September 1994.

[6] Texas Instruments, "TMS320C6201/6701 Evaluation Module," SPRU269D, December 1998, p. 2-9.

[7] Texas Instruments, "Analog Applications Journal—New DSP Development Environment Includes Data Converter Plug-Ins," August 2000.

[8] http://www.ti.com/sc/docs/tools/mso/dataconv/plug_in.htm.

11

Engineering Design Assistance Tools

11.1 Introduction

Designing functional blocks in a system can be relatively straightforward; however, the majority of the engineering effort for any system is involved in efficiently implementing the functions inside the blocks and integrating the blocks together to form a working system. This is true for hardware-based radio systems, where ASIC development is emphasized, as well as software-based systems, where the implementation and integration effort is mostly the domain of the DSP and software engineer. To improve the ability for product developments to meet an ever-growing need for faster time to market, engineering design assistance (EDA) tools have emerged over the past ten years. They aim to decrease the number of errors introduced during the design and implementation stages and increase the rate of bug fixes during the later implementation and integration stage.

11.2 How EDA Tools Can Help Software Radio Development

EDA tools provide the means to simulate a complete system on a nontarget hardware platform (usually a UNIX or WINPC machine) and allow the software engineer to explore a portion of the system in detail. For 3G software radio these simulations are mostly non-real-time, and speed can be an important issue when applied to large complex simulations. EDA tools are designed to rapidly capture problems (often graphically), analyze them, and provide the engineer with easy to interpret output data (e.g., graphs and displays).

Depending on the ability of the tool, the simulation can be performed at a high level (behavioral or functional), with progressively lower levels of abstraction, right through to simulations of hardware, where cycle-accurate models are constructed. For systems where the target hardware (e.g., DSP) is fundamentally different from the simulation machine, a cycle accurate model requires the target hardware instruction set simulator (ISS) to be integrated into the EDA tool (Figure 11.1).

Tools can be offered with a suite of optional libraries containing predeveloped algorithmic blocks ready for use. Complete reference design kits (RDKs) for wireless air interfaces are now becoming widely available. RDKs are often built using standard library functions and provide a baseband simulation of a complete up- or downlink, including the transmission channel. In the example shown in Figure 11.1, a system is simulated using the preverified library functions; system outputs are captured for a set of input test vectors. The library function, "Function2.c," is then replaced by a DSP implementation, "Function2.asm," and the process is repeated to test the functionality and performance of the DSP version by comparing the results of the simulation with those produced by the library functions. The DSP algorithm can then be tuned (reduced cycle and memory consumption) for maximum performance while checking that overall system performance is maintained.

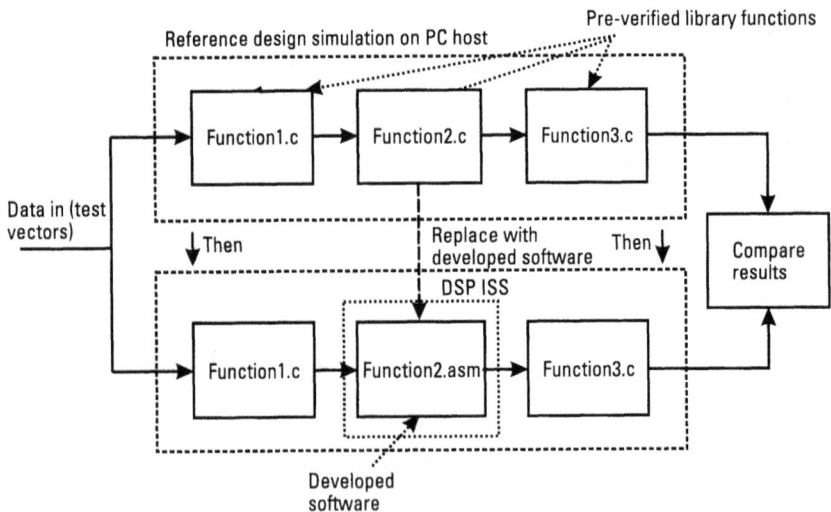

Figure 11.1 EDA simulation example.

Some EDA tools are suited to exploring a system's data path only (Figure 11.1), while others can also simulate the control plane by including state machine modeling functionality. Each of these features helps the engineer to isolate a problem area without needing the real system.

11.3 MATLAB

Matrix Laboratory, or MATLAB, started as a software modeling tool in 1984; it operates on both PC and UNIX computers. At its core the program uses matrix manipulation to solve mathematical and scientific problems. Algorithms are programmed using a C-like scripting language:

```
% SPREADING CODE FOR SECOND CHANNEL
  % if the spreading code has not been assigned
  if length(nCode) < 2
      C = [ones(1,nCode / 2) -ones(1,nCode/2)];
          % for(i = 1:nCode)  C(i) = (-1)^i; end;
      else  % if the spreading code has already been assigned
      C = nCode;
      nCode = length(C);
  end;
```

MATLAB includes extensive graphical display features for presenting problem results. Additional libraries are available to speed the development of simulations; the communications, filter design, and signal processing toolboxes are examples.

Users who wish to avoid learning the MATLAB language can extend the tool by adding Simulink. Models can be built in a GUI by selecting graphical blocks from a library and creating data paths by simply joining wires. Using this capability users no longer have to describe their problem as a matrix, and the visual representation of the algorithm can simplify verification. The following Simulink blocksets (libraries) are available: communications, DSP, fixed point, C6701 developer's kit, and CDMA reference.

11.3.1 C6701 Developers Kit

The developer's kit for TI DSP is a part of Simulink that provides a rapid way of targeting algorithms to the Texas Instruments' C6701 EVM; see Chapter 10 for more EVM details. The kit allows the developer to generate code to download to the EVM's DSP and run as a standalone system.

Simulink blocks are provided to interface to the EVM's on-board audio analog to digital and digital to analog converters; this is only suitable for implementing audio applications. Links are also provided to Code Composer by using the real-time data exchange (RTDX) mechanism; this allows the EVM to interact with Simulink and speed simulations by running code that would otherwise be run by the host PC.

11.4   Cocentric System Studio

Cocentric System Studio (CCS) is a UNIX-based EDA tool developed by Synopsys; the product is an upgrade and replacement for COSSAP. The product is described [1] as a SystemC simulator and specification environment for joint verification and analysis of algorithms, hardware, and software at multiple levels of abstraction. Reference libraries include the CDMA2000 1xRTT, WCDMA/UMTS-FDD, EDGE, TD-SCDMA, GSM/GPRS, IS 136, and IS-95 standards. One advantage of CCS over the previous COSSAP is its support for state machines and control plane simulation.

11.5   SPW

Signal Processing Worksystem (SPW) is also a UNIX EDA tool [2] used for data flow–based algorithmic design, analysis, and implementation. Key features of the tool include WCDMA/UMTS, CDMA2000, GSM, IS-95, IS 136, and adaptive antenna library models. Hardware and software cosimulation is possible, including cosimulation of a system environment where the DSP software implementation uses a DSP instruction set.

11.6   SystemC

The SystemC language is being developed by the Open SystemC Initiative (OSI) [3] as a de facto open source standard for system-level design. Source code is freely available for download from the OSI Web site. This open source approach is designed to allow EDA vendors to build tools based on SystemC and promote interoperability between vendors' products. The language architecture [4] uses C++ as its base and adds functionality by overlaying a core language, new data types, elementary channels, standard channels, and methodology-specific channels.

11.7 Reference Design Examples

All of the tools covered in this chapter offer cellular mobile wireless RDKs suitable for software radio development.

11.7.1 CDMA Reference Block Set

MATLAB's Simulink can be extended by adding the IS-95A reference library (i.e., the CDMA reference block set). The library includes base station receiver, base station transmitter, mobile station receiver, mobile station transmitter, and common blocks. There are several demonstration systems built using the library blocks, including a complete forward link, as depicted in Figure 11.2.

The forward traffic channel end to end model consists of the base station transmitter, transmission channel, and mobile receiver. This is a baseband model, where the imperfections associated with the up and down-conversion (between baseband and RF) functions have been ignored for simplicity. The conversion functions would normally include a stage of digital frequency conversion followed by a stage of analog frequency conversion. The channel model is a combination of Rayleigh multipath fading and additive white Gaussian noise (AWGN). The model measures the bit error and frame error rates for various parameter combinations. Figure 11.3 illustrates the results when the signal to noise ratio (Eb/No) for the channel model is varied. The number of errors is calculated for ten frames of transmitted data and, hence, the zero bit error rate for Eb/No –2 dB (i.e., insufficient data to yield a statistically meaningful result).

11.7.2 NIST

The National Institute of Standards and Technology (NIST) [5] has developed a work program to help the cellular communications industry test the wideband CDMA components of the IMT2000 family of standards (i.e., CDMA2000 and UMTS/WCDMA). NIST has worked with Cadence to produce a free downloadable set of CDMA2000 [6] simulation models for the SPW tool. Two UMTS/WCDMA simulation models are also available from the NIST Web site [7], the first is written in C and the second is implemented using Simulink.

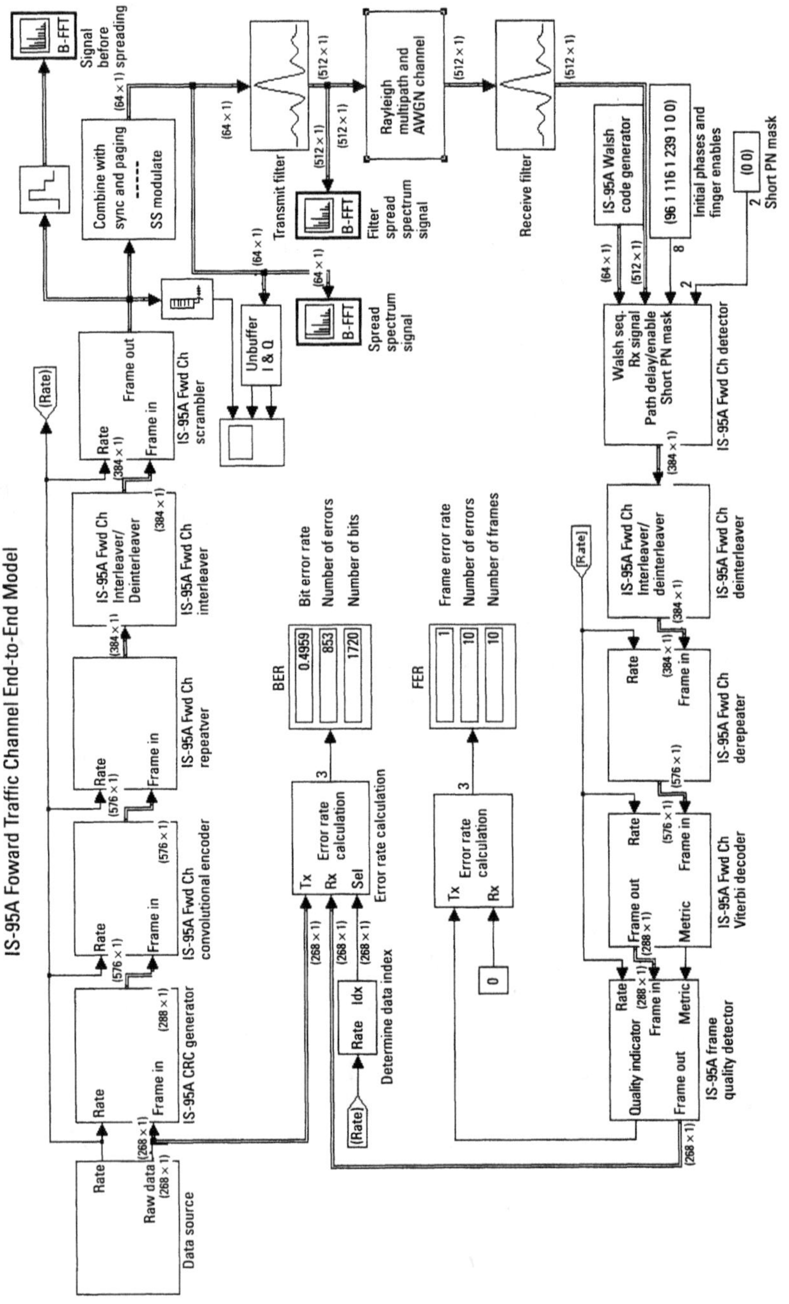

IS-95A Foward Traffic Channel End-to-End Model

B-FFT

Signal before spreading

Combine with sync and paging

SS modulate

(64×1)

Transmit filter

(512×1)

(64×1)

(512×1)

(512×1)

Rayleigh multipath and AWGN channel

(512×1)

B-FFT

Filter spread spectrum signal

Receive filter

(512×1)

(64×1)

IS-95A Walsh code generator

$(96\ 1\ 116\ 1\ 239\ 1\ 0\ 0)$

Initial phases and finger enables

$(0\ 0)$

Short PN mask

(64×1)

(512×1)

8

2

2

Walsh seq.
Rx signal
Path delay/enable
Short PN mask

IS-95A Fwd Ch detector

(384×1)

Unbuffer I & Q

(64×1)

B-FFT

Spread spectrum signal

Frame out

Rate

(384×1)

Frame in

IS-95A Fwd Ch scrambler

(Rate)

IS-95A Fwd Ch Interleaver/ Deinterleaver

(384×1)

IS-95A Fwd Ch interleaver

Rate

(576×1)

Frame in

IS-95A Fwd Ch repeater

(384×1)

Bit error rate
Number of errors
Number of bits

BER

| 0.4959 |
| 853 |
| 1720 |

Frame error rate
Number of errors
Number of frames

FER

| 1 |
| 10 |
| 10 |

IS-95A Fwd Ch Interleaver/ deinterleaver

(384×1)

IS-95A Fwd Ch deinterleaver

Rate

(384×1)

Frame in

IS-95A Fwd Ch derepeater

(576×1)

Rate

(576×1)

Rate

(576×1)

Frame in

IS-95A Fwd Ch convolutional encoder

Rate

(576×1)

Frame in

3

Tx
Rx
Sel

Error rate calculation

(268×1)

(268×1)

(268×1)

Error rate calculation

3

Tx
Rx

Error rate calculation

Frame in

IS-95A Fwd Ch Viterbi decoder

Frame out
Metric

(288×1)

Frame in
Metric

Rate

(576×1)

Rate Idx

Determine data index

(Rate)

Rate

Raw data

(268×1)

Frame in

(268×1)

IS-95A CRC generator

Rate

Data source

0

Quality indicator

Rate

(288×1)

Frame in

(288×1)

Frame out

IS-95A frame quality detector

(268×1)

(Rate)

Figure 11.2 IS-95A forward traffic channel model.

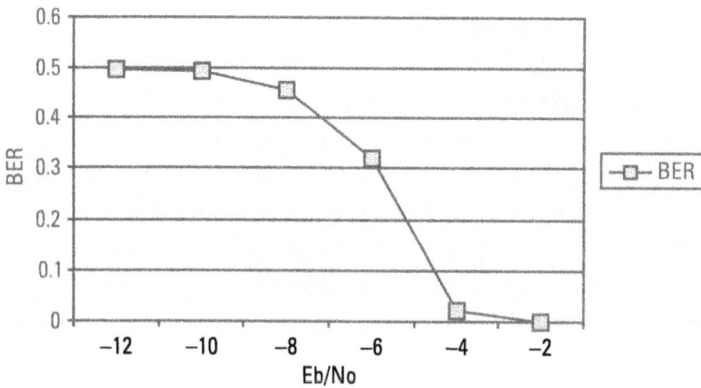

Figure 11.3 IS-95A BER versus Eb/No.

11.8   Conclusion

When engineers first started debugging DSP software in the 1980s without EDA, the only solution was to build a simulation tool specifically for the job, in many cases using now outdated languages such as BASIC and FORTRAN. As the complexity of wireless systems grow the availability, features, and cost of commercially available EDA tools will continue to increase. Interestingly, EDA tools play their part in the exponential rise of complexity by helping to complete larger and larger projects in similar time frames. Each subsequent project tends to place a larger demand on the tools that respond to meet the challenge, and so the cycle is perpetuated.

References

[1] Synopsys, "Cocentric System Studio Data Sheet," 2002.

[2] http://www.cadence.com/whitepapers/wireless_white_paper.html, "Obstacles to 3G Wireless and Design Tools and Methods to Overcome Them," 2001.

[3] http://www.systemc.org.

[4] Swan, S., "An Introduction to System Level Modeling in SystemC 2.0," Open SystemC Initiative, http://www.systemc.org/projects/sitedocs/documents/SystemC_WP20/en/1, May 2001.

[5] http://w3.antd.nist.gov/.

[6] http://w3.antd.nist.gov/wctg/3G/CDMA 2000_form.html.

[7] http://w3.antd.nist.gov/wctg/3G/wcdma/.

List of Acronyms

1G	First-Generation Cellular Mobile Phone Systems
2G	Second-Generation Cellular Mobile Phone Systems
3G	Third-Generation Cellular Mobile Phone Systems
3GPP	Third-Generation Partnership Program
3GPP2	Third-Generation Partnership Program 2
4G	Fourth-Generation Cellular Mobile Phone Systems
Abis	Interface between a BTS and a BSC
ac	Alternating Current
ACM	Adaptive Computing Machine
ADC	Analog-to-Digital Converter
ALU	Arithmetic Logic Unit
AM	Amplitude Modulation
AMPS	Advanced Mobile Phone System (1G Standard)
API	Application Programming Interface
ARIB	Association of Radio Industries and Businesses
ASIC	Application-Specific Integrated Circuit
ASSP	Application-Specific Standard Part
AWGN	Additive White Gaussian Noise

BSC	Base Station Controller (2G Term = RNC in UMTS)
BTS	Base Transceiver Station (2G Term = Node B in UMTS)
CCD	Charge Coupled Device
CDMA	Code Division Multiple Access
CDMA2000	A 3G standard administered by 3GPP2
CDMA2000-1x	Any variant of CDMA2000 that is limited to a 1.25-MHz channel
CDMA2000-3x	Any variant of CDMA2000 that is limited to a 3.75-MHz channel
CDMA2000-3x-MC	Multicarrier version of CDMA2000-3x
CDMA2000-3x-DS	Direct-sequence version of CDMA2000-3x
CISC	Complex Instruction Set Computer
CORBA	Common Object Request Broker Architecture
COTS	Commercial off the Shelf
CPU	Central Processing Unit
CRC	Cyclic Redundancy Check
CW	Continuous Wave
CWTS	China Wireless Telecommunications Standards Group
DAC	Digital-to-Analog Converter
dB	Decibel
dBFS	Decibel relative to Full Scale
dc	Direct Current
DDC	Digital Downconverter
DMA	Direct Memory Access
DNL	Differential Nonlinearity
DOA	Direction of Arrival
DPR	Dual-Port RAM
DSP	Digital Signal Processor

DUC	Digital Upconverter
EDA	Engineering Design Assistance
EDGE	Enhanced Data rates for GSM Evolution
EMC	Electromagnetic Compatibility
EMI	Electromagnetic Interference
EMIF	Enhanced Memory Interface
ETSI	European Telecommunications Standards Institute
FCC	Federal Communications Commission
FDD	Frequency Division Duplex
FDM	Frequency Division Multiplexing
FFT	Fast Fourier Transform
FIFO	First In First Out
FIR	Finite Impulse Response
FM	Frequency Modulation
FOM	Figure of Merit
FPGA	Field Programmable Gate Array
GMSK	Gaussian Minimum Shift Keying
GPIO	General-Purpose Input/Output
GPRS	General Packet Radio Service
GPS	Global Positioning System
GSM	Groupe Speciale Mobile or Global System for Mobile communications
HDL	Hardware Description Language
HDR	Hardware Defined Radio
HF	High Frequency (3–30 MHz)
HPI	Host Port Interface
IBM	International Business Machines
IDE	Integrated Development Environment
IEEE	Institute for Electrical and Electronics Engineers
IIR	Infinite Impulse Response

INL	Integral Nonlinearity
IS95	Interim Standard 95 (equivalent to CDMA One or narrowband CDMA)
ISS	Instruction Set Simulator
ITU	International Telecommunication Union
LFSR	Linear Feedback Shift Register
LNA	Low Noise Amplifier
LRU	Line Replaceable Unit
LSB	Least Significant Bit
MAC	Multiply and Accumulate
McBSP	Multichannel Buffered Serial Port
Mcps	Mega Chips per Second
MIPS	Millions of Instructions per Second
MMAC	Million Multiply and Accumulations
MMACS	MMAC per second
MMSE	Minimum Mean Square Error
MSB	Most Significant Bit
MSC	Mobile Switching Center
MSPS	Mega Samples per Second
NCO	Numerically Controlled Oscillator
NMT	Nordic Mobile Telephone (1G analog standard)
Node B	UMTS term for base transceiver station
NOP	No Operation
NRE	Nonrecurrent Engineering
OAM	Operations Administration and Maintenance
OMG	Object Management Group
OS	Operating System (e.g., LINUX, Windows)
OSF	Oversampling Factor
OVSF	Orthogonal Variable-rate Spreading Factor
PA	Power Amplifier

PC	Personal Computer
PCI	Peripheral Component Interconnect
PLL	Phased Locked Loop
PMR	Private Mobile Radio
QASK	Quadrature Amplitude Shift Keying
QCELP	Qualcomm Code Excited Linear Predictive
QoS	Quality of Service
QPSK	Quadrature Phase Shift Keying
RAM	Random Access Memory
RCP	Reconfigurable Communications Processor
RF	Radio Frequency
RISC	Reduced Instruction Set Computing
rms	Root Mean Square
RNC	Radio Network Controller (UMTS term for BSC)
ROM	Read Only Memory
SAW	Surface Acoustic Wave
SBC	Single Board Computer
SDE	Software Development Environment
SDR	Software Defined Radio
SDRAM	Synchronous Dynamic Random Access Memory
SDRF	Software Defined Radio Forum
SFDR	Spurious Free Dynamic Range
SIMD	Simultaneous Instruction Multiple Data
SINAD	Signal to Noise and Distortion Ratio
SLOC	Source Lines of Code
SNR	Signal to Noise Ratio
SoC	System on a Chip
SSRAM	Synchronous Static Random Access Memory
TDD	Time Division Duplex
TDM	Time Division Multiplex

TDMA	Time Division Multiple Access
TIA	Telecommunications Industry Association
TTA	Telecommunication Technology Association
TTC	Telecommunication Technology Committee
UE	User Equipment
UHF	Ultra High Frequency (300–3,000 MHz)
UMTS	Universal Mobile Telecommunication Services
U.S.	United States
USB	Ultra Serial Bus
UTC	Universal Time Coordinated
UTRAN	UMTS Terrestrial Radio Access Network
VGA	Variable Gain Amplifier
VHDL	VHSIC Hardware Description Language
VHF	Very High Frequency (30–300 MHz)
VHSIC	Very High Speed Integrated Circuit
VIM	Velocity Interface Mezzanine
VITA	VMEBus International Trade Association
VLIW	Very Long Instruction Word
VME	VERSAmodule European
WARC	World Administrative Radio Conference
WCDMA	Wideband CDMA (5 MHz for 3G)

About the Author

Paul Burns is currently a 3G manager with NEC in Melbourne, Australia, and is working on the company's tri-mode GSM/GPRS/WCDMA terminal development project. From 1999 to 2002 he played a pivotal role with the SDR start-up Advanced Communications Technologies Australia (ACT). As project manager at ACT, Mr. Burns was responsible for all engineering efforts dedicated to the SpectruCell project. SpectruCell is a software defined cellular mobile base transceiver station (BTS) that is capable of handling multiple 2G and 3G air-interfaces. The engineering team completed the world's first dual air-interface GSM and IS95B software-defined BTS in 2001. The SpectruCell project has also resulted in several patents and the development of a complete in-house BTS framework, including hardware, middleware, and application software. Mr. Burns also operates Simplexity Communications (http://www.simplexity.com.au) a telecommunications software consulting company, which specializes in SDR. He also plans to introduce a new industry forum to further advance software defined radio technology.

Mr. Burns completed his electronics engineering degree at the University of South Australia in 1986. He started his telecommunications and signal processing career at GEC-Marconi in the United Kingdom, where he successfully developed a suite of TMS320 assembly language DSP algorithms to test the EH101 sonar signal processing platform. Following this, Mr. Burns moved to CSA in Kent, England, where he designed military and broadcasting RF systems and delivered projects to Plessey and the BBC. In 1990, Mr. Burns returned to Australia and continued in communications systems with Vicom. As product manager, he was given significantly increased responsibilities and served on the company board as a director. At Vicom, he took a lead role in the communications systems division and successfully delivered a number of large broadcasting systems to Telstra. In 1993, Mr. Burns moved to Telstra to commence work on the billion-dollar Jindalee Over The Horizon Radar Network (JORN). At JORN, he led a team of radar system integration and test engineers, and presented a paper on radar network testing during the 1995 ITEA conference in New Mexico. He took a major role during the handover of the project to Lockheed Martin/ Tenix during 1997, including a major redesign and reorganization. In April 2001, Mr. Burns presented at 3G Technical Strategies in Geneva on optimum base transceiver/node B architectures.

In his spare time Mr. Burns enjoys traveling, snow skiing, and wine tasting with his wife Josephine.

Index

Recent Titles in the Artech House Mobile Communications Series

John Walker, Series Editor

WCDMA: Towards IP Mobility and Mobile Internet, Tero Ojanperä and Ramjee Prasad, editors

Wireless Communications in Developing Countries: Cellular and Satellite Systems, Rachael E. Schwartz

Wireless Intelligent Networking, Gerry Christensen, Paul G. Florack, and Robert Duncan

Wireless LAN Standards and Applications, Asunción Santamaría and Francisco J. López-Hernández, editors

Wireless Technician's Handbook, Andrew Miceli

For further information on these and other Artech House titles, including previously considered out-of-print books now available through our In-Print-Forever® (IPF®) program, contact:

Artech House	Artech House
685 Canton Street	46 Gillingham Street
Norwood, MA 02062	London SW1V 1AH UK
Phone: 781-769-9750	Phone: +44 (0)20 7596-8750
Fax: 781-769-6334	Fax: +44 (0)20 7630-0166
e-mail: artech@artechhouse.com	e-mail: artech-uk@artechhouse.com

Find us on the World Wide Web at:
www.artechhouse.com

www.ingramcontent.com/pod-product-compliance
Lightning Source LLC
Chambersburg PA
CBHW050455190326
41458CB00005B/1287